SpringerBriefs on Case Studies of Sustainable Development

Series editors

Asit K. Biswas, Third World Centre for Water Management, Los Clubes, Atizapan, Mexico
Cecilia Tortajada, Los Clubes, Atizapan, Mexico

The importance of sustainable development has been realized for at least 60 years, even though the vast majority of people erroneously think this concept originated with the Brundtland Commission report of 1987 on Our Common Future. In spite of at least six decades of existence, we only have some idea as to what is NOT sustainable development rather than what is. SpringerBriefs on Case Studies of Sustainable Development identify outstanding cases of truly successful sustainable development from different parts of the world and analyze enabling environments in depth to understand why they became so successful. The case studies will come from the works of public sector, private sector and/or civil society. These analyses could be used in other parts of the world with appropriate modifications to account for different prevailing conditions, as well as text books in universities for graduate courses on this topic. The series of short monographs focuses on case studies of sustainable development bridging between environmental responsibility, social cohesion, and economic efficiency. Featuring compact volumes of 50 to 125 pages (approx. 20,000–70,000 words), the series covers a wide range of content—from professional to academic—related to sustainable development. Members of the Editorial Advisory Board: Mark Kramer, Founder and Managing Director, FSG, Boston, MA, USA Bernard Yeung, Dean, NUS Business School, Singapore.

More information about this series at http://www.springer.com/series/11889

Jean-Paul Close
Editor

AiREAS: Sustainocracy for a Healthy City

Phase 3: Civilian Participation—Including the Global Health Deal Proposition

OPEN

Editor
Jean-Paul Close
AiREAS
Eindhoven
The Netherlands

ISSN 2196-7830   ISSN 2196-7849   (electronic)
SpringerBriefs on Case Studies of Sustainable Development
ISBN 978-3-319-45619-5   ISBN 978-3-319-45620-1   (eBook)
DOI 10.1007/978-3-319-45620-1

Library of Congress Control Number: 2016950235

This Springer imprint is published by Springer Nature
The registered company is Springer International Publishing AG
The registered company address is: Gewerbestrasse 11, 6330 Cham, Switzerland

Foreword

In the interest of solving many of our societal problems (e.g., air pollution, sustainability, health issues), technological innovations alone are not enough: Human behavior also needs to change. Research into persuasive technology (Fogg 2003; IJsselsteijn et al. 2006) investigates how we can use the technology that people interact with while making the relevant decisions (e.g., in regard to their car) needed to influence human behavior or thinking. This relatively young research area promises to provide solutions and deliver the insights needed to change the human behavior related to our societal problems. At the same time, this area still needs study to uncover many of the basic mechanisms of how technology can help people change and adapt (see also, Midden and Ham 2012: Persuasive technology to promote pro-environmental behavior). One of the open questions is in regard to the crossroads of usability and persuasion: What are the interactions between the usability of technology and the persuasiveness of technology? A highly motivated team of students (Joyce Brouns, Tim van den Boom, Marjan Hagelaars, Relinde van Loo and Daniëlle Ramp) set themselves up to analyze this interaction in a very societally relevant application domain: The newly developed AiREAS App, which is intended to provide users with information about the air quality in their direct surroundings as measured by the AiREAS measurement network in the city of Eindhoven, The Netherlands. This app may contain a variety of persuasive strategies (e.g., providing information about air pollution levels). Still, usability problems with the app might diminish its persuasive power, and even more complex interactions between usability and persuasiveness might be identified. The report on this research can be read in the following pages and is one of the first steps into this important new domain. These are important issues, as much from a societal perspective as a scientific perspective. The full power of persuasive technology (in helping to solve current societal issues) can only be unleashed when we better understand this kind of technology, and are able to build systems that are both usable and persuasive.

Dr. Jaap Ham

Preface

One of the biggest complexities we encounter when desiring the transformation or evolution of a reality is the way people relate to a reigning culture or paradigm through a particular mindset. For them, this perception of a current reality gives comfort and security. It is intensely difficult when this mindset needs to be revised voluntarily for the sake of sustainable human progress. What is sustainable human progress? Why would anyone want to participate in a transformation? With what authority can we ask people to look at their reality in a different way? Why would anyone let go of a lifestyle to adopt a new one?

My own breakthroughs in awareness have been described in depth in Chap. 2 of the book on Phase 1 as manifesting in the origin of STIR and subsequently AiREAS. This new state of mind enabled me to look at our reality in a totally new way. I saw the complex duality of our current human existence on planet Earth:

- Recognize the human being as a member of a very smart and creative species, a kind of miracle of evolution of life, capable of developing tremendous tools to help in its own wellness; a unique species that has come so far that it hardly has to fear shortages or problems if it organizes itself well
- Recognize this same human being as the sixth cause of massive elimination of life-forms, including our own, since the birth of our planet!

When this duality manifests itself in one's own awareness, a choice appears. Do I address my own wit and creative potential to deal with life and the sustainable evolution of the species? Or do I blindly continue being part of this destructive reality into which we have evolved, peaking during the last few decades up to unsustainable heights?

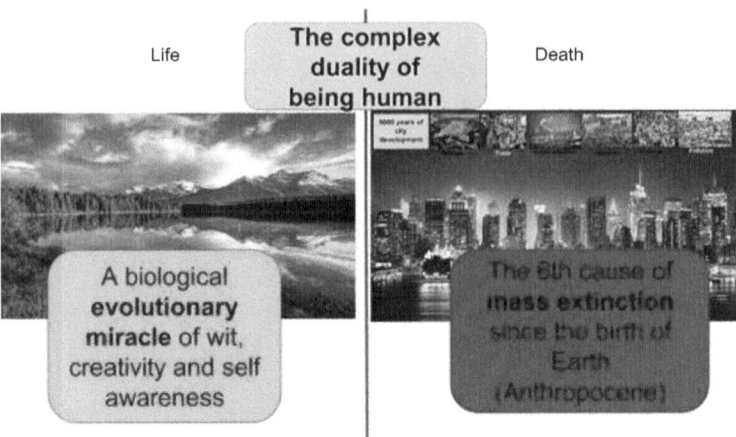

We all have this choice. When we make it, a whole new reality appears. AiREAS invites us to make this choice for health and healthy air

The choice seems simple, but the consequences are huge. Our global community is using its wit mostly for destructive self-interests, being part of the problem rather than the solution. During my opening speech at the global announcement of AiREAS in October 2012, I asked my fellow citizens to take responsibility for our core values by joining me and others in AiREAS, creating a movement from destruction to sustainable human progress (Sustainocracy). The next speech was delivered by Eindhoven's alderman, Mary-Ann Schreurs, who related to my words "taking responsibility." She said that one also has to "*be able* to take responsibility," referring not just to awareness, but also to the many impediments people encounter when they want to make the choice. It is not just the people who encounter difficulties; the institutions and their executives also encounter huge challenges and discouraging situations that inhibit them from moving forward when awareness strikes. Our POP research was therefore much more challenging, because we needed to determine ways in which individuals and executive people of the institutions could find comfort in taking responsibility and discover ways around the impediments.

Our investigation went to great length to interact openly with people when inviting them to look at their own lifestyles, daily choices, and voluntary changes when confronted with information about health, air quality, and exposure. We are still confronted with a society that, in essence, rewards polluting activities out of old political, social, and economic interests. For instance, people commute between home and salaried work in fossil fuel-consuming cars. We organize logistics involving diesel trucks around our consumption patterns. We destroy our landscapes for massive manipulated food production. The pharmaceutical industry pollutes the water, air, and soil with medicines, pesticides, and fertilizers. We have developed a throwaway lifestyle around (plastic) packaging. In essence, our money-driven political and economic reality is blindly perverse when dealing with real core values. Our financial systems were designed to reward fragmented, industrialized, and commercial processes, taxing the population and enterprises to organize remedial activities through bureaucracy and social services. We are not financially structured to reward "proactively doing good" from a social or environmental point of view. While things such as a house, food, recreation, and health care are financed through income obtained by "contributing to the problem," "doing good" is seen as a common idealistic activity.

People often comment on their understanding of pending and worrying issues, but also the impossibility of their taking responsibility due to an overruling societal format revolving around money: "Yes I know, but I need the money to pay my mortgage." This duality was significant for our research. What comes first? A new societal format with which to resonate, one which delivers the necessary securities to its participants in a new way? Or a change in mentality for people that have the guts, or simply have no choice but to distance themselves from the old format in order to start developing the new?

The answer is both! We need a choice and we need a need to choose. And even then the choices are often still subject to a hierarchy of priorities. It is not the choice itself that matters so much, it is what one lets go of in the process of making the choice. There is therefore a considerable mental effort involved in the process of making choices. Such effort translates into all kinds of reaction, from stress to satisfaction, from fear to passion, and from reluctance to excitement. In essence, we distinguish between the blind pursuit to "have what we need" and trusting our wit to "(co)create what we need." Those who create tend to evolve, while those who just consume tend to die mentally in reluctance, apathy, and greed, even before dying physically.

This is easily said, but the mainstream of a community is captured through the organized structure of perceived and desired possession, even if everything we have or obtain is accessed through establishing a financial or natural debt. This debt, if expressed in money, is against the economy; if expressed in natural values, it is against nature, life, and our evolution. Research had shown that up to 36 % of the local population in Eindhoven was worried about air pollution and the possible effects on their own health. Recent scientific reports provide worrying insights of children being born with behavioral or physical disorders and reduced life expectancy, enhancing such senses of discomfort due to air pollution. Most people, however, still hold government responsible for producing corrective measures, failing to see themselves as a significant part of the problem and hence also of the solution. Often, public responsibility for pollution is blamed on traffic, combustion of fossil fuels, and industrial activities. The lack of available alternatives and the perceived need for a car and other material consumables makes people reluctant to challenge their own lifestyles. The lack of awareness is high, even though worries are significant. Air pollution, in general, is invisible, and hence does not immediately belong to the daily reality that we perceive and take into account. Thanks to this project, we were able to look at and describe the real situation in regard to pollution, health, and lifestyle and come to new insights. We also had a chance to glance at cultural and demographic differentiation, giving rise to strong desires to further elaborate on this in future programs, in Eindhoven and across the world.

Using phase 1, the ILM

The ILM project of AiREAS, phase 1, published through Springer at the beginning of 2016, was designed to make visible the invisible at the level of human exposure to air pollution as it affected citizens and visitors in their outdoor activities in town. This fixed network of airboxes is designed to measure local conditions in the vicinity of elder homes, schools, and city quarters where people reside, and shopping places where people tend to gather or commute in different ways. Scientists and government officials helped decide the position of each airbox in order to get the best theoretical insight into the pollution that surrounds each measurement station. Thirty five airboxes deliver a full set of measured information to a database every 10 min. This database has open access, allowing us to pick up and use the near real-time data for communication purposes.

The ILM measures:

Particulate Matter:	PM 1, 2.5 and 10
Ultrafine Dust:	UFP (6 out of the 35 units)
Gases:	Ozone, NO_2
Environmental data:	temperature and humidity
Technical data:	GPS coordinates, time of the day

The historic database is kept for interpretation and analysis by both scientists and government officials. This infrastructure was the basis for defining the healthy city project that was to follow and involve 4000 citizens through health and lifestyle research, using the health data for interpretation and communication, together with the database of air pollution. Due to the enormous complexity (medical, technological, social, team effort, interpretation, etc.) of this project, a smaller Proof of Principle, or POP for short, was set up first, with just 40 citizens. In Chap. 2, we will go into the details of this process and the interaction we engaged in with the citizens involved.

Persuasive interaction with citizens became a challenge on its own. Included in phase 1 (air quality data gathering) was a small budget for communication. We had considered the 3D approach presented by the University of Madrid to show layers of air pollution through a 3D animation over the town. We also looked at integrating light codes into the city's streetlight system. But at this stage, it was too far-fetched to consider integrating systems. Instead, we decided to set up a website (http://www.aireas.com) with the intention of getting interaction going with the citizens through this means. This worked to a limited degree at best.

A website provides highly informative data for those who look for it. Accessing a website indicates a pre-existing level of interest in the topic covered by the site. In the case of air pollution and health, the people who visit the site still need to make up their own minds about what they see and how they interpret the mix of data without the benefit of specialized knowhow. Little interaction was achieved, even though the site served its informative purpose perfectly, especially for a minority of committed citizens.

Mobile App for instant information

Our network partner Imtech (later Axians) had suggested making an App for mobile phones. Instead of making it a co-creation effort within the multidisciplinary context of AiREAS, they decided to develop it as an internal experiment with a commercial motivation behind it. The ILM network was released as a measurement system in December 2013. The mobile App was ready in May 2014. It was shared with a very small amount of people for testing.

The availability of the App was decisive in getting instant access to the network and near real time data anywhere simply through a mobile phone. It instantly revealed high peaks through color coding of the sensors, displayed as dots on a map of the town.

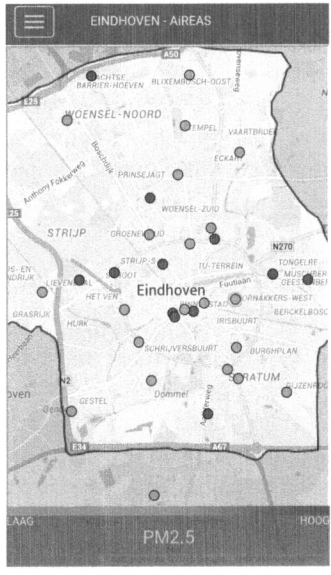

The App for mobile phones; each dot is an Airbox, coded like a traffic light (red = bad, orange = risky, green = good) (This type of coding information was changed over the course of 2015 through interaction with the Ministry of Health (RIVM). Discussion addressed the thresholds when changing color and the unjust psychological effect of 'green' as safe.)

This allowed us, the initiators of the AiREAS healthy city movement, to document cases of pollution when alerted by the app. The mobile access to real time information, in the case of an alarming event, would trigger instant field research through human observation and dialogue, allowing for a combination of local civilian observation with the measurements of the system, weather information and specialist interaction that would together determine the details of the case. Within 24 h, such dynamic interaction would provide us with insights about curiosities detected in our own city. The invisible had become visible and highly alive.

For first line research, the App was a good tool, especially for those active in and committed to the field of observation and analysis. But the question arose as to whether the App would serve the interests of the larger public. Would people want to pay for the App? What use would they get out of it? And how would the App help influence social behavior? These questions needed specific research.

The University of Technology in Eindhoven has a department led by Dr. Jaap Ham. It teaches *persuasive technology*. This specialized field looks at the ways in which technology influences the behavior of people. An interesting next step triggered by AiREAS was to see how the objective of behavioral change could be proactively achieved through applied technology, knowledge, and communication skills. The field of "persuasive communication" was accessed and drew the enthusiastic attention of Dr. Ham. During a minor, a short course of a few months for students who wish to enhance or broaden their education with other fields of interest, about the subject of persuasive communication, students from many different fields of study would get acquainted with the concept of value-driven persuasion. They were asked to analyze the AiREAS App from that perspective and come up with suggestions for optimum persuasion.

The progression of our experience had accelerated, transforming from basic useless marketing attempts, or the regulation of pollution, to a persuasive invitation for general involvement in proactive health and air quality development through innovation. The peer 4 "participation society" or "awareness-driven eco-system" had started, even if we ourselves did not yet know it when we started.

Seven Chapters and the Global Health Deal:

In Chap. 1, we will go into great detail about the AiREAS App analysis through persuasive communication skills, using a case study on Jaap Ham's activities by one of his groups of students.

By January 2015, the POP had started, as the result of which we attempted to involve 40 local civilians in a medical and lifestyle research trajectory. This proved much more challenging than anticipated. The choices we made and the experiences we gathered are described in Chap. 2.

In June 2015, our technological partner TNO expressed a desire to find 12 volunteers to participate in a backpack experiment for the measurement of direct exposure to air pollution. The backpack was filled with measurement and GPS equipment and would be carried for 5 days, 24 h a day, both indoors and outdoors, by each participant. AiREAS was an ideal partner to set this up professionally, and the information could enhance our own POP research on lifestyle. All that we learned can be read in Chap. 3.

Persuasion has many fields of attention, and one of the POP objectives was to see if we could stimulate or activate the innovative drive among the population. When people relate to a new reality, then new productivity can appear that may translate into new forms of entrepreneurship, the development of services and products and even a new element in the economic reality, referred to as the Transformation Economy. In Chap. 4, we explore this incubational drive and the early results that we obtained.

Similarly, we wanted to address the multicultural reality of our city. We wanted to obtain the involvement of many subcultures, of which the biggest are Turkish and Moroccan. We had entered into an exchange program for students with AiREAS's sister enterprise STIR Academy, a peer 4 initiative for the development of participative learning. The European Erasmus+ subsidy program was brought in by our partner Stichting BdT, bridging the Turkish and Dutch cultures. Seven thousand students from all over Turkey are to visit us in Eindhoven over the coming years. By the end of 2015, we had already received 700 students and some 50 teachers. We involved them all in the AiREAS Healthy City challenge with admirable results. These will be described in Chap. 5.

Finally, we decided to experiment in combining our ILM and POP research with particular events in Eindhoven. It was announced that the Marathon of Eindhoven would become one with innovative components proper to the characteristics of the city. Sport in the town's streets, training, health, and air quality match nicely with an experiment in communication and human interaction. Over 200,000 individuals would support the 23,000 participants during that Sunday in October 2015. Such an audience is a good platform for experiments. Chapter 6 will describe to you what we did and what we learned.

Summing up in Chap. 7, we will relate all of the above to the definition of a peer 4 regional development and the important consequences of the multidisciplinary, core-driven innovative partnership between the local government and its population. This evolutionary step at a regional level is new. To consolidate the method, we need to define very well how it functions and show the results to the world. It is a strong response to all of the key issues of destruction of our habitat and health that we face everywhere in the world, worth the while of local leadership to adopt such evolutionary steps to strengthen local harmony, economy, and resilience. We will also see that we learned much more than expected, about our behavior, our lifestyles, our adaptiveness, and our interpretation of such health and health hazards as pollution.

The 7 chapters will hopefully serve as a source of inspiration for other communities, enhancing our experiments, learning processes, and incubative innovations with those of their own in their own regional context. Every region is culturally and demographically unique and different, producing a diversity of challenges and innovations that can surprise the world. Adopting peer 4 sustainocratic processes has its obvious advantages, not just in generating a new economic evolution, but also in empowering regional leadership. We are ready to help anyone get started and consolidate such decisions.

To finalize the book, we publish the executive summary in **the annex**, in which we use the evidence obtained to make a call for a **Global Health Deal**. This establishes the context of all dialogues and socioeconomic development around the core values that can make our global community and human evolution robust and sustainably progressive.

Eindhoven, The Netherlands   Jean-Paul Close

Contents

Introduction

In 2011, we[1] founded the sustainocratic[2] AiREAS cooperation with the objective of taking responsibility as a community for the co-creation of healthy regions, using air quality, human health, and regional dynamics as measureable guiding principles. With local health as our leading drive, the cooperation was uniquely set up through the direct multidisciplinary involvement of the four pillars of society:

- local citizens,
- local government,
- creative entrepreneurial innovators,
- scientists/educators.

This way of sharing responsibility around core values of human existence, of which Health is one example, had been previously defined and settled in the STIR Foundation. STIR was established in 2009 in Eindhoven to (re)define human complexities for sustainable human progress through value-driven experimentation. It found its first practical application in the context of health related to air quality and human dynamics. The resulting project-driven type of multidisciplinary co-creation was given the formal name of AiREAS. The method of regional co-creation, driven by core values, was coined Sustainocracy,[3] representing a new way of addressing our democracy by accepting that our core values should lead the way in defining innovative community priorities. In 2015, external researchers[4] recognized the practical execution of AiREAS Eindhoven as a peer 4 regional development, an evolutionary step representing an awareness-driven eco-system of

---

[1]AiREAS' founders are Jean-Paul Close and Marco van Lochem.

[2]"Sustainocratic" represents a multidisciplinary way of working on natural core human values defined in Sustainocracy.

[3]*Sustainocratie, de nieuwe democratie waarin de mens centraal staat*—October 2012/MultiLibris.

[4]Venture Spring report about Smart City projects in Eindhoven—October 2015.

value-driven co-creation.[5] At the same time, the cooperation received the European VINCI Energy Innovation award[6] as a group and individually in the category of partnerships.

The vision, experimental steps, theoretical referencing with other scientific sources, and the achievements that sprang from all of these are being published worldwide through the channels of Springer[7] and New Horizons.[8]

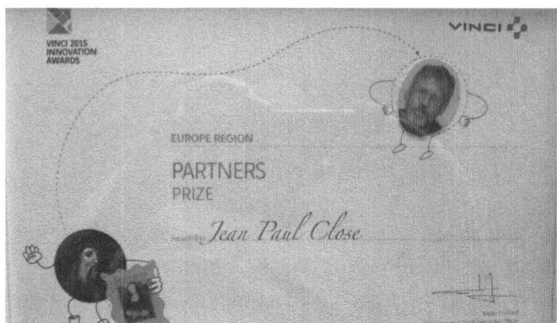

A copy of the Vinci Award

Citizen Involvement

This particular book describes the experience obtained in the field of spurring citizen involvement and participation in taking responsibility for regional core values such as health and air quality, representing a new, innovative societal context[9] for all involved. The research done shows the difficulty for human beings in relating to new paradigms, no matter how important those paradigms may be for them, especially when surrounded by a dominant socioeconomic reality that does not necessarily relate directly to core values. It also describes the challenge that this type of regional development contains for the entire community, its structures, leadership, socioeconomic context, and the way citizens interact with each other, with governance and with the environment.

The research was conducted in 2015 and the first months of 2016 in combination with phase 1 and 2 of the AiREAS healthy regional development in Eindhoven. These first two phases were as follows:

---

[5]Level 4 global co-creation defined by the Presencing Institute.

[6]https://www.vinci.com/vinci.nsf/en/newsroom/pages/collective_intelligence_supporting_urban_air_quality.htm.

[7]The Spiritual Dimension of Business Ethics and Sustainability—Springer 2015.

[8]Sociology and Anthropology 3(6): 311–317, 2015, Redefining human complexities, New Horizons.

[9]Context differentiation between money dependence and co-creation for core values.

- Phase 1: Making visible the invisible[10] (published through Springer in 2016) through a fine maze real-time measurement infrastructure in Eindhoven for air quality and human exposure, referred to as the ILM,
- Phase 2: Linking air quality with human health and lifestyle, referred to as the POP (Proof of Principle). The POP evolved into two fields of investigation:
  - Noninvasive medical research and lifestyle investigation/POP1
  - Lifestyle and human persuasion for lifestyle innovation/POP2.

We decided to publish the two fields of investigation of the POP separately, as each contains its own field of expertise, interest, and newly acquired insights. So, POP2 (civilian participation) became phase 3, while POP1 (medical research) was named phase 2. The preliminary phases (ILM and POP1) were necessary to provide toolkit data and knowledge as input for professional interpretation and open public communication. The key objective was to address the population with the invitation to innovate our lifestyles and daily instruments of use to improve our health and environment.

Complex Team Effort

When we started the Proof of Principle, we had the desired end results in mind but no idea as to how we would get there. The civilian participation part was especially filled with desires and expectations, but also a total lack of knowledge. We needed to build up experience along the way. Each step would become decisive in regard to the next ones that would need to be taken. Whatever we planned was always challenged by a different outcome than expected. Adapting, changing, or adjusting our approach continuously became part of a co-creation routine that gave us insight for enlargement of the programme in Eindhoven, as well as other regions. This adaptive flexibility was also the basis for the outcome, which proved to be much more diverse and complete than could have been anticipated.

Phase 3, hence, has provided us with unique know-how about and insights into human behavior, influences, and participation in processes of change. We also learned a lot about the way surroundings influence the behavior of people, personal leadership, perception, and the effect of awareness on people as to whether or not to take initiative. We learned about human motivation, persuasion, passive and active innovation, responsibility, and the stimulus of human productivity, including entrepreneurship. The living lab setting was our own city of Eindhoven and its surroundings. A whole new reality revealed itself to us, together with the complexity of dealing with change and value-driven innovation.

---

[10]http://link.springer.com/chapter/10.1007/978-3-319-26940-5_3.

The team that conducted this particular POP2 research consisted of:

- Nicolette Meeder—personal interviews and feedback gathering
- John Schmeitz—database development, analysis, and reporting
- Jean-Paul Close—team coordinator and overall researcher of behavior, entrepreneurship, and change.

We had the expert support of three sources of information and feedback:

- The open data from the ILM near real-time air quality measurement system
- The POP1 medical research team, consisting of Dr. Eric de Groot and Dr. Pierre Cluitmans
- Jaap Ham—assistant professor at the Technical University of Eindhoven for persuasive technologies and communication.

The following research environments in Eindhoven were all eventually used during the course of 2015 to get persuasive interaction with civilians, students, researchers, innovation, and technology:

- The phase 1[11] mobile App experiment, based on the ILM
- The phase 2 proof of principle of health and lifestyle research—POP1
- Hackathon/creativity with open data
- Participative education through the STIR Academy with Erasmus+
- Multicultural addressing of subgroups in the city
- The backpack lifestyle research
- The marathon event in Eindhoven.

Each of these environments brought their own unique context and experiences, which are described in separate chapters. Altogether, they provided a solid foundation for peer 4 regional development and stimulus of local productivity, civilian participation, and fundamental changes in the functioning of the city. The project also gives us insight into a community that arises around new common values such as health and environmental awareness, and how to deal with the securities that maintain cohesion and progress when money is no longer the driving force.

The method used for this research was therefore:

- Use the core values of health, air quality, and lifestyle as the dot on the horizon
- Consider the entire city's population to be partners in the healthy city objective
- Connect a multidisciplinary team to the persuasion for improvement, using local research as a tool for triggering innovation
- Find ways to connect with the complexity of a fully operational society to gain as much insight as possible into behavior, consequences, and support for a paradigm shift toward caring for health (proactive) rather than trusting health care (reactive)

---

[11]AiREAS phase 1, making visible the invisible, Springer 2016.

- Determine how to connect with the population to enhance such a paradigm shift using positive feedback, innovation, and communication
- Dynamically cluster our partners around the experimental environments that we encounter along the way and which would eventually determine our own learning curves.

Within this framework, we could experiment with an open mind, analyzing our progress through forward driven initiatives and backward interpretation. Our learning curve about citizen participation within the overall abstraction of developing core human values was goal using air quality, human health, and human dynamics as quantifiable instruments. We had the expected and financial resources, now we needed to define scenarios to do our experiments within the fully operational society.

References

Fogg BJ (2003) Persuasive technology: using computers to change what we think we do. ISBN-13: 978-1558606432

Ijsselsteijn WA et al (2006) Virtual fitness: stimulating exercise behavior through media technology. Presence-Teleop Virt Environ 15(6): 688–698

Midden C, Ham J (2012) Persuasive technology http://link.springer.com/chapter/10.1007/978-3-642-31037-9_8

Chapter 1
Persuasive Communication

Jean-Paul Close and Jaap Ham

1.1 Introduction

In marketing terms, a business model uses communication instruments to attract customers or develop brand awareness and loyalty. There is always a degree of manipulation in such communication, as we try to influence human purchasing decisions through persuasion. Persuasion is therefore more than the simple act of providing information; it is a professional and measureable expertise in influencing decision-making. In commerce, there is a competitive environment in which a company tries to excel. When, however, we deal with the introduction and deployment of peer 4 regional development, we are not selling anything. We are inviting the population to take active part in the development of local core human and natural values that affect us all. This participation can be measured in the amount of response achieved, changes observed and improvements measured within the pursuit of the core values desired. In AiREAS, we measure air quality, health and regional dynamics. This gives us a great responsibility when dealing with persuasion. We don't lie or twist the truth to get a sale; we <u>positively</u> invite for innovative change with validated information and interpretation of data. And we measure and celebrate our results in terms of health and air quality development, not our bottom line financial gain.

J.-P. Close (✉)
STIR Foundation/AiREAS, Sustainocracy, Eindhoven, The Netherlands
e-mail: jp@stadvanmorgen.com

J. Ham
Department of Technology, Management Section and Human Technology,
Eindhoven University of Texchnology, and Tech4Change, Eindhoven, The Netherlands
e-mail: j.r.c.ham@tue.nl

© The Author(s) 2016
J.-P. Close (ed.), *AiREAS: Sustainocracy for a Healthy City*,
SpringerBriefs on Case Studies of Sustainable Development,
DOI 10.1007/978-3-319-45620-1_1

The ILM provides us with near real time information about the fine maze status of air quality in the city of Eindhoven. The experimental App for mobile phones, developed by Imtech/Axians,[1] provided the same information as the AiREAS website, together with a color change for each airbox when certain values were exceeded. Figure 1.2 above represents a screenshot of the App. At the level of AiREAS's founders, we made effective use of the App to observe the network and register specific cases of pollution as they occurred in real time. When we noticed something odd through the App, for instance, a sudden peak in one or more airboxes, we could go into more depth by opening a special PC application developed by our Airbox co-creation manufacturer and partner ECN. All of these large, experimental, tools were only in the hands of partners within the kernel of AiREAS. The question arose as to whether these tools were suitable for extensive public use for the persuasive effect of participation and social innovation? We ourselves were, of course, already motivated; otherwise, we would not have started such a complex mission as AiREAS. But could we motivate others?

1.2   University for Technology[2] in Eindhoven (TUe)

Different universities participate in AiREAS, each from their specific area of expertise through highly motivated and socially committed specialists. Results that they achieve in AiREAS are documented and published, specifically by their individual academic partners, but also together from a group process perspective. The ILM and academic partnerships also led to new research, majors, minors, Ph.D. and master programs, etc. The University of Twente, for instance, uses the AiREAS infrastructure and partnership organization for their DAMAST project, financed through Maps4Society.[3] DAMAST is an automated risk assessment program on exposure to air pollution. At the same time, the university professors and researchers have been very influential in the spatial design and rollout of the ILM—phase 1. Involvement, therefore, has a two-way reciprocal value for AiREAS, the university, the students and the scientists.

In the same sense, Dr. Jaap Ham had already been involved in early stages of AiREAS for the purpose of influencing the 'soft' side of our activities, referring to human interaction and behavioral analysis, as opposed to the 'hard' side consisting of applied technology, economy, products and infrastructures. This 'soft' side is highly underestimated when dealing with issues of sustainable progress. When dealing with the financing of new AiREAS projects, attention tends to go to the tangible, material implications that define economies, rather the intangible,

---

[1]https://www.axians.nl/over-axians/vinci-energies/.

[2]https://www.tue.nl/en/.

[3]Maps4Society is an initiative of the Dutch Technology Foundation http://www.stw.nl/nl/programmas/maps4society.

immaterial ones that define social wellness, co-creation and involvement. In AiREAS, we were introducing a new balance by trying to make the invisible visible, not just in air quality but also in human behavior, motivation and change. Jaap had already surprised the group when the discussion turned to the possible importance of 'awareness' in provoking change. I had introduced the concept of civilian BAGE,[4] a staged process of people who become aware of the need for change, accept the responsibility to do so and get rewarded afterwards. Jaap reflected back that no awareness is needed. People are essentially followers of the pack. They tend to do what their neighbors do or blindly behave according to the cultural patterns defined locally, no questions asked. This is referred to as "the principle of least effort".[5] A daily routine is always 'normal' unless something happens that upsets that routine. It is human nature then to try to get back to 'normal'. This hardly involves the revolution of disruptive change; it is mostly a safe status quo or gradual evolution.

Jaap used the example of the enormous marketing effort in town to introduce solar panels through the distribution of flyers in areas with an extensive amount of home ownership. No commercial successes were registered at all, despite local awareness about the need to address energy issues. That is, until a young entrepreneur visited his uncle in town. The youngster had started a business installing solar panels and asked his uncle if he could place some on his roof. The uncle accepted and even offered to have a word with a friendly neighbor to see if he also wanted to participate. The solar panels were installed, and within 6 months, the entire street was using them.

This example showed the tremendous waste of resources involved in the distribution of flyers while the personalized approach had far-reaching and much more cost-effective consequences. Between the youngster and his uncle, there was a level of trust, just like the relationship between the uncle and the befriended neighbor. The rest was just copycat behavior of a street that does not want to stay behind. This learning process also became significant in the subsequent processes of AiREAS. We learned that personal attention and interaction generated a much larger and more positive response to change than any other means of communication. Persuasion needed to be personalized, based on trust and reciprocal personalized reward. When persuasion achieves its objective in a small portion of society and becomes visible, then human nature will do the rest.

Awareness on its own is therefore not enough to influence the masses. Many people are aware of pollution, climate change or the amount of plastic in the sea. They understand the problem, but barely translate that into a need for individual behavioral modification. 'Why should I change if the rest of the world does not?' The follower mentality is also culture-driven. 'Everyone does it, so I do the same, even if I am aware that it is wrong or should be done differently.' Within this

---

[4]"Sustainocratie, de nieuwe democratie waarin de mens centraal staat", in Dutch (2012, MultiLibris).

[5]Zipf, George Kingsley. "Human behavior and the principle of least effort." (1949).

spectrum, the concept of persuasion becomes interesting. How can early adopters, the pioneers of change, finally produce the entire chain result as if it were a line of dominos falling after the first one is pushed over? How do we involve such early adopters? What instruments persuade? How do we visualize the positive contributions? It was Henry Ford, automobile manufacturer, who said, about disruptive innovation, "If I had asked the masses about their innovative wishes for mobility they would have asked for a faster horse".

Change is hence a leadership issue, not a democratic or market working process, in which leadership sees a new reality well before it is commonly present in the mainstream activities.

With this in mind, we looked again at our mobile App. The tool was clearly informative and useful for the selected group of AiREAS intimi. But was it persuasive? With this question, we entered the expertise of Jaap Ham, who decided to use it as a learning tool in his short course on persuasive communication.

1.3   Method Used

Every year, students from all fields of study within the TUe can take a short course of 2 months created by Jaap. They can choose the cases they want to work out and, as such, enter a participative learning process. One of the options to choose from is the AiREAS App.

When a group chooses to take on the App, they receive a presentation from AiREAS Sustainocrats about the objectives, the background of the App, and the challenge of persuasion that we face. They also receive access to usage of the App. Then, they have 2 months to develop their views, applying scientific knowledge provided by Jaap and the world's literature on the subject of persuasion.

The first group of students to reflect on the App did so purely from a point of view of persuasion without additional concrete adaptations. Subsequently, they would add something new, showing an evolution in the processing of the tooling. The tone was set with a score of 3 out of 10 for the App on the scale of persuasion. This is, of course, low in an absolute sense, but for AiREAS, a valuable key indicator. We need to recognize that the App was never primarily envisaged from a point of view of persuasion and had been experimentally developed by technicians who wanted to display the information of our measured data on a handheld device. Having a working model of such an App allowed us to engage in the learning process of this new field of communication, together with the creativity of the students. A 3 out of 10 triggered a lot of curiosity among the AiREAS partners and opened up the dialogue between value-driven entrepreneurship, persuasion, citizenship and technology.

The students were asked to produce a report, a presentation and a poster. One of those reports excelled and received an 8.5 out of 10 from Jaap. AiREAS was also enthusiastic about the insights developed by these students and decided to incorporate their entire work (with their permission, of course) into this publication.

They hence appear as co-authors of this chapter. The report is complete enough to finalize this chapter with its publication. It is now up to AiREAS and its expansion across the world as to how we shall use this knowledge in pursuit of our health deal with the world.

The report

0HAUB0—USE-HIT-3 Projects—The Human in Technology—Research Project Report—2014–2015

USE - HIT Projects 2014 - 2015

Research Project Report

Analyzing the persuasiveness and usability of the AiREAS application in raising awareness of air pollution and decreasing the contribution of the individual

Supervisor: Jaap Ham

Authors:

Joyce Brouns,
Tim van den Boom,
Marjan Hagelaars,
Relinde van Loo,
Daniëlle Ramp,

Introduction
Air pollution is an increasing problem in our current day society. Polluted air can cause cardiovascular and lung diseases, which especially affect old people, children and people already suffering from such diseases.[6]

---

[6]"Effecten." [Online]. Available: http://www.rivm.nl/Onderwerpen/F/Fijn_stof/Effecten (Accessed: 07-May-2015).

This research will mainly focus on particle pollution. Particle pollution is the term for a mixture of solid particles and liquid droplets found in the air, such as dust, dirt or smoke. Particle pollution[7] consists of inhalable particles, which are categorized by size:

- Ultrafine particles with a diameter smaller than 0.1 μm (PM 0.1)
- Fine particles with a diameter around 2.5 μm (PM 2.5)
- Respirable suspended particles with a diameter around 10 μm (PM 10).

The smaller the particles, the more danger they pose. Small particles can reach deep into the lungs after inhalation or even pass the vessel walls and get into the bloodstream.[8]

Recent studies have shown that an increase in air pollution created an increase in cardiovascular and lung patients. The inhalation of these particles can cause inflammatory responses and toxic effects. Lung cancer and lung fibrosis are especially common diseases in polluted areas, and children suffer more from asthmatic attacks.

Particle pollution is produced through two separate processes: mechanical and chemical. There are different sources of particle pollution. On average, industry is responsible for 41.1 % of particle pollution, mobility for 28.9 and households for 7.7 %, as indicated in Fig. 1.1. To decrease particle pollution, various measures can be taken, including pollution prevention, control technologies and control measures, and may be implemented through regulatory, market-based or voluntary programs.[9] Decreasing the contribution of individuals or households to air pollution is another important measure that can improve the air quality. Changing everyday behavior can have an immediately positive effect, for example, using bicycles instead of automobiles or dressing warm instead of using the fireplace. To change behavior, awareness of the production of pollution has to be raised. Here is where technology comes in. Technology can be used as a persuasive tool to change behavior by raising awareness.

One example of a persuasive technology related to fine particle pollution is the AiREAS application. In general, AiREAS is an organization that wants to raise awareness about air pollution and achieve their goal of a cleaner and healthier city. The function of the application itself is to use the data to notify people of the different polluted zones. This can persuade people in two ways: if the zone is highly polluted, individuals can choose to avoid polluting activities, such as not using the barbeque or the fireplace. Also, people will be able to avoid highly polluted areas when travelling through the city.

---

[7]O. US EPA, "Health|Particulate Matter|Air and Radiation|US EPA." [Online]. Available: http://www.epa.gov/airquality/particlepollution/health.html (Accessed: 07-May-2015).

[8]O. US EPA, "Basic Information|Particulate Matter|Air and Radiation|US EPA." [Online]. Available: http://www.epa.gov/airquality/particlepollution/basic.html (Accessed: 07-May-2015).

[9]Cees Midden and Jaap Ham, Persuasive technology to promote environmental behaviour, vol. Hoofdstuk 23.

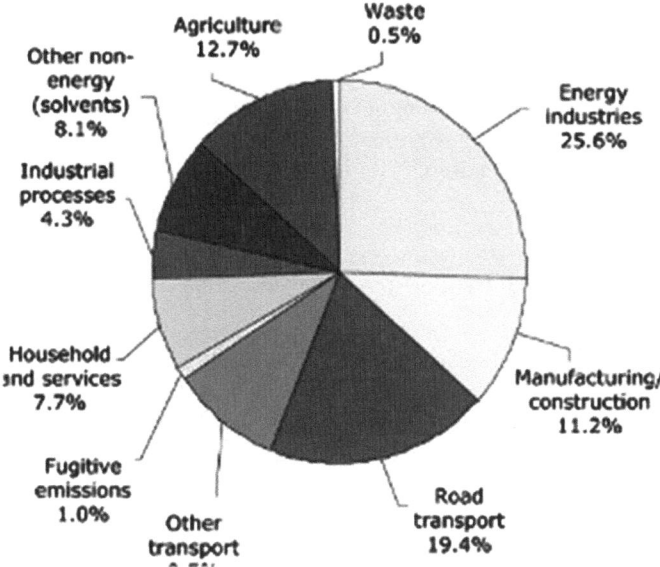

Fig. 1.1 Sources responsible for particle pollution in Europe (SurveyMonkey, 2 juni 2015, https://www.surveymonkey.com/)

However, using this kind of technology can only raise awareness if the application is actually persuasive and useable enough. Therefore, the current research analyzes the persuasiveness and usability of the AiREAS applicationand, where necessary, provides new ideas for improving the application.

1.4 Research Question 1

What strategies of persuasiveness based upon Fogg, Oinas-Kukkonen and HarjumaaandCialdini are already present in the AiREAS application?

1.5 Method

1.5.1 Participants and Design

In order to test the persuasiveness already present in the AiREAS application, three different persuasive strategy frameworks were used, e.g., Fogg, Cialdini and

Oinas-Kukkonen and Harjumaa, as previously described. We have evaluated the application in regard to every aspect of each strategy with a score between 1 and 10. This score will be based on the degree of presence of a certain technique in the current application, with 1 meaning it is not present in the current application and 10 meaning that it absolutely is present. The average score of every aspect of each strategy is shown in Appendix I.

1.6   Materials and Procedure

All criteria were applied according to the literature of Fogg,[10] Cialdini[11] and Oinas-Kukkonen[12] and Harjumaa. We have independently examined the degree of presence of the different criteria of each framework and scored these. The five different scores of every criterion were averaged, and thereafter, an average end score of each framework was calculated. These results were processed in a bar graph and will be compared with each other to determine the persuasiveness of the application.

1.7   Results

1.7.1   Hypothesis Testing

Before analysis was performed, we took a brief look at the application in order to get an idea of its persuasiveness. It was concluded that the application mostly provides information about the fine dust distribution in Eindhoven, but it is not, in fact, very persuasive. Therefore, we hypothesized that the application will get a low score on persuasiveness according to the three different frameworks of Fogg, Cialdini and Oinas-Kukkonen and Harjumaa. In order to test this hypothesis, five users examined the different criteria of each framework and scored these as described above. These scores were averaged in order to determine the persuasiveness of the current application.

---

[10]Fogg, B. J. (2002). Persuasive technology: Using computers to change what we think and do. Ubiquity, 2002 (December), 5.

[11]Cialdini, R. B. (1993). Influence: The psychology of persuasion. New York, Morrow.

[12]Oinas-Kukkonen, H., & Harjumaa, M. (2008). A systematic framework for designing and evaluating persuasive systems. In Persuasive technology (pp. 164–176). Springer Berlin Heidelberg.

Fig. 1.2 Results of the persuasiveness analysis according to the three different frameworks of Fogg, Cialdini and Oinas-Kukkonen and Harjumaa

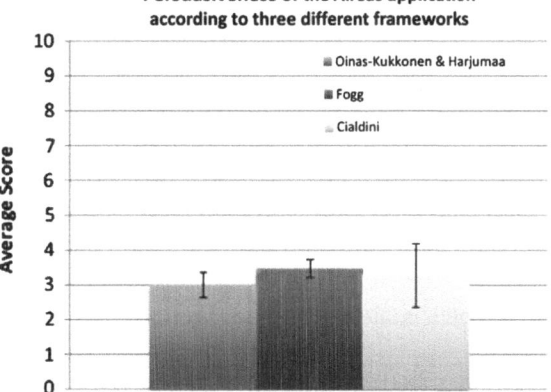

1.7.2   Exploratory Analysis

The results are shown in Fig. 1.2. Raw data can be found in Appendix I. The graph in Fig. 1.2 shows the three average scores of the three different frameworks. It can be seen that all of the frameworks have an average score around 3. However, the application scores the highest on persuasiveness according to Fogg and the lowest according to Oinas-Kukkonen and Harjumaa.

1.8   Conclusion and Discussion

The persuasiveness of the AiREAS application has been examined using the criteria of Fogg, Caldini and Oinas-Kukkonen and Harjumaa. Five different evaluators have independently executed the persuasiveness of the application. The five different scores of every criterion were averaged, and thereafter, an average end score of each framework was calculated. The results can be seen in Fig. 1.2. All the frameworks received an average score around 3, which means that the AiREAS application is not very persuasive. Based on Cialdini, the application is the least persuasive, and based on Fogg, the application is the most persuasive. Because of the low scores, it can be concluded that many persuasive strategies are lacking in the current application. In order to reach the goal of persuading users to produce less fine dust or avoid polluted areas, far more persuasive strategies have to be implemented in the AiREAS application. One of the most important of these is probably the possibility of tracking the user's own behavior, so as to make a personal application that is different for each user. As people can be sensitive to the accomplishments of others and may try to top them, other important strategies that can be implemented in the application are social comparison, competition and social facilitation. When users can track other users' behavior as well, competition is possible.

Furthermore, people tend to be more easily persuaded into acting upon certain behavior when there is a chance of being rewarded, so virtual rewards would also be a good addition to the AiREAS application. In research question 3, more detailed research is provided in which the type of feedback best suited for creating behavioral change is investigated.

1.9 Research Question 2

How usable is the AiREAS application?

1.10 Method

1.10.1 Participants and Design

Usability can be defined as the extent to which a product can be used by specified users to achieve specified goals with effectiveness, efficiency and satisfaction in a specified context of use. We evaluated the usability of the AiREAS application ourselves by analyzing the following five primary criteria: effectiveness, efficiency, learnability, memorability and satisfaction.

After we evaluated the AiREAS application ourselves, we asked five random participants from different backgrounds and age groups to evaluate the application. The only thing they had in common was that none of them had ever used the AiREAS application before.

1.11 Materials

We scored the above five criteria on a scale of 1–10. By averaging these scores, we determined a usability score for the AiREAS application. We provided the participants with the System Usability Scale (SUS).[13] SUS is a quick, reliable method for measuring the usability of a system. The SUS questionnaire consists of ten items with five response options to choose from, ranging from 'strongly agree' to 'strongly disagree'.

---

[13]A. S. for P. Affairs, "System Usability Scale (SUS)," 06-Sep-2013. [Online]. Available: http://www.usability.gov/how-to-and-tools/methods/system-usability-scale.html (Accessed: 07-May-2015).

1.12   Procedure

We used and analyzed the application separately, before evaluating it on its usability by using a table to score the criteria. We approached the participants personally and let them use the AiREAS application for a couple of minutes. We didn't tell them the purpose of the evaluation up front, so that the usability would be evaluated in an unbiased manner. After the participants indicated that they had seen the whole application, they were provided with the SUS to evaluate the system on its usability.

1.13   Results

1.13.1   Hypothesis Testing

After quickly browsing through the application, we came to expect that it would prove not to be very usable, since it lacked certain basic features. This hypothesis applied to the SUS questionnaire as well, as the participants evaluated the same system; it could be assumed that they would miss the same features as we did.

1.13.2   Exploratory Analysis

In our own analysis, and based on the 1–10 evaluation of the assorted criteria, the application scored best on efficiency and worst on satisfaction, as can be seen in Fig. 1.3. The scores were averaged, which gave an outcome of 4.7. The outcome of the SUS questionnaire was higher than that of our own evaluation. With a score of

Fig. 1.3 Evaluation of AiREAS application on five criteria

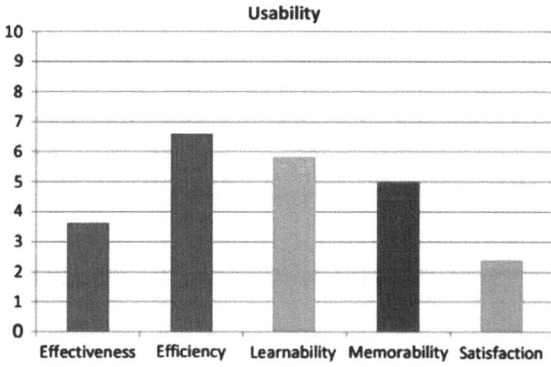

Fig. 1.4 The average
outcome of our own
evaluation versus the
evaluation of our participants

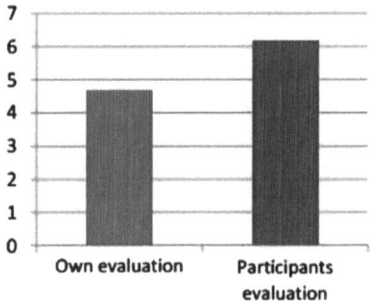

62 on a scale of 100, the participants evaluated the AiREAS application better than
we did, as can be seen in Fig. 1.4. The raw data are shown in Appendices II and III.

1.14   Conclusion and Discussion

In averaging these scores (4.7), we agreed that the application was not very satis-
fyingly usable. The average outcome of the evaluation of the other participants was
higher than that of our own evaluation. We assume this was due to our bias, since
we approached the application with a more critical eye, given that we are
researching it. The most striking result was that all participants found the application cumbersome to use and that they all assumed that they would not tend to use
the application with any real frequency. We assume that these results are not
directly related to the usability of the application, but rather that they are related to
the lack of a clear goal.

We believe that the AiREAS application has high potential, since it could
contribute to the solution of a major societal issue, namely air pollution. AiREAS
could be used to make users more aware of their behavior in regard to air pollution,
but we found that our participants were not really aware of this. The main
improvements should therefore focus on making it more attractive for the user to
use the application frequently and making the goal and relevance of the application
more clear. Future research could be done in both these fields.

1.15   Research Question 3

Does direct feedback in the form of red/green colors or verbal feedback in the
AiREAS application improve the awareness of humans regarding the presence of
fine dust?

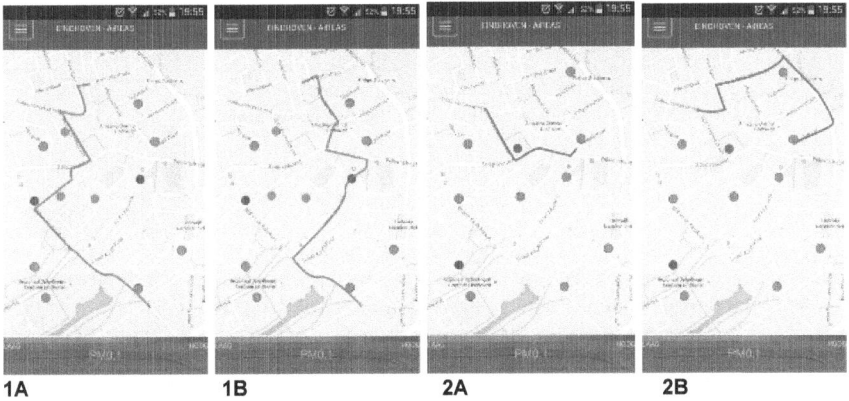

1A   **1B**   **2A**   **2B**

Fig. 1.5 Picture 1**A** and 1**B** show two routes with the same length; the participant had to choose either **A** or **B**. Picture 2**A** and 2**B** show two routes for which the length differs and the healthier route is longer; again, the participant had to choose either **A** or **B**

1.16 Method

1.16.1 Participants and Design

In order to answer this research question, an online questionnaire was created in three different versions. One version contained no feedback, one contained direct verbal feedback and one had direct feedback with red- and green-colored symbols. We wanted to represent the city of Eindhoven, because the AiREAS application is only available in that city. There are currently 223,220 people living in Eindhoven,[14] so, using a 95 % confidence interval and a 10 % margin of error, 68 participants were needed in order to achieve reliable results.[15] Eventually, 75 participants submitted the questionnaires, giving us a margin of error of 9.3 %.

1.17 Materials and Procedure

The questionnaire consisted of ten questions each, with a choice between two different routes. These routes differed from each other in their amount of pollution, and the participants had to choose which route they would take. An example of these choices can be seen in Fig. 1.5. In questions 4, 7 and 8, the choice was more difficult for the participants, because the unhealthy choice was the shorter route. This way, the influence of time over health could be evaluated.

---

[14]Eindhoven Buurtmonitor, 1 januari 2015, http://eindhoven.buurtmonitor.nl/.

[15]SurveyMonkey, 2 juni 2015, https://www.surveymonkey.com/.

Table 1.1 Example of the different kinds of color and verbal feedback in the questionnaire

	Verbal feedback	Colour feedback
Correct	Well Done! This is the most healthy route. Although this route guides you through a polluted area, the alternative route guides you through more polluted areas	
Wrong	Too bad! This is not the most healthy route. This route guides you through a lot of highly polluted points, while the alternative route guides you through only a single polluted point	

Three different versions of the questionnaire were used, to test the influence of different kinds of direct feedback based on the experiment by Midden.[16] Questionnaires featuring direct verbal feedback and feedback in the form of red/green colors were tested against the questionnaire that contained no feedback.

The direct verbal feedback consisted of an explanation as to why the answer was either correctly or wrongfully chosen. The feedback in the form of red/green colors showed a green thumbs up when the answer was correctly chosen or a red thumbs down when the answer was wrongfully chosen. An example of the feedback is given in Table 1.1.

1.18   Results

1.18.1   Hypothesis Testing

The verbal feedback explains why the answer is wrong or correct, and our expectation was that this would increase understanding of how to interpret the two images. The participants with color feedback would know if they had gotten a question wrong or correct, however they would not know the reason. Their learning curve would increase, however to a much lesser degree than the learning curve of the participants with verbal feedback. Participants without feedback had no

---

[16]Cees Midden and Jaap Ham, *Persuasive technology to promote environmental behaviour*, vol. Chap. 23.

understanding of whether their answer was correct. They would not learn anything from the questionnaire. This gives us the expectation that the participants with verbal feedback would answer the highest number of questions correctly and the percentages of correctly answered questions would rise, while the participants without feedback would answer the lowest number of questions correctly. When answering the question as to which route is shorter than the other, most of the participants would answer this question correctly, while the persons without feedback or with color feedback would not choose the right answer.

1.19 Exploratory Analysis

The results are given in Fig. 1.6, and the raw data can be seen in Appendix V. The vertical axis shows the percentage of participants that answered the question correctly and the horizontal axis shows the specific questions. The participants with verbal feedback answered the greatest number of questions correctly. Also, questions 4, 7 and 8, in which the choice was given between the shorter, less healthy route and the longer, healthier route, were better answered by the participants with verbal feedback, while the persons who received no feedback or color feedback mostly chose the unhealthy option.

In Fig. 1.7, the number of participants who answered the question correctly is shown for questions 4, 7 and 8. The learning curve of the participants without feedback is higher than the learning curve of the participants with verbal feedback. It can be seen that the participants with direct verbal feedback answered the questions better overall. However, the participants with no feedback have a higher learning curve.

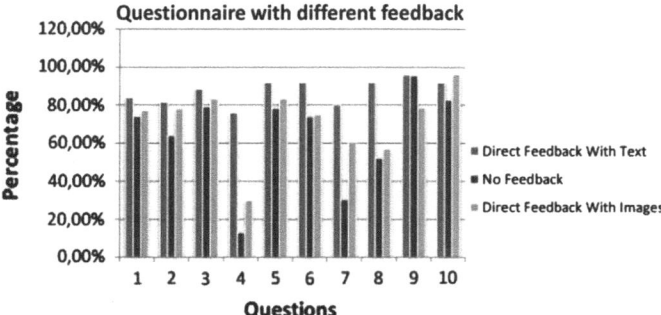

Fig. 1.6 Bar graph of the results of the questionnaires

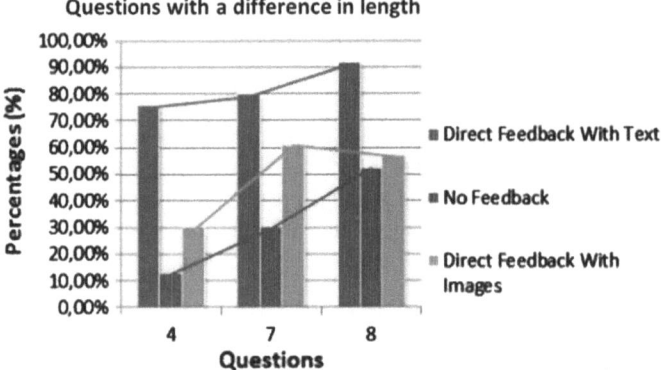

Fig. 1.7 Bar graph of the results of questions 4, 7 and 8

1.20 Conclusion and Discussion

In this research question, we tried to investigate whether direct feedback in the form of red/green colors or verbal feedback in the AiREAS application improves the awareness of humans regarding the presence of fine dust. In order to do so, we created a questionnaire with ten questions; the participants had to choose between two routes. This questionnaire had three different versions, one with direct verbal feedback, one with direct color feedback and one without any feedback at all. In the results of this research, we found that the participants with the direct verbal feedback faired best on the test.

These findings show us that giving feedback indeed changes the behavior of people and can improve it in this way. Direct verbal feedback made a particularly significant difference in this test. The participants with direct color feedback and no feedback provided unhealthier answers. However, they did show a decent learning curve when given situations in which the less healthy route was shorter. These findings could help us to better understand human behavior and could perhaps help other designers in trying to improve the environmental awareness of humans.

In our research, we used an online questionnaire. This way, unfortunately, we were unable to be sure if participants only answered the questionnaire once. We also did not tell the participants whether they were riding a bike, walking or driving a car. Participants may have chosen differently if this had been explained. These limitations were not taken into account when processing the results. However, we do not think that these limitations changed the result of our research.

In conclusion, when trying to improve the awareness of humans regarding the presence of fine dust, using direct verbal feedback is a good option.

1.21 Overall Conclusion and Discussion

Particle pollution is a huge problem in current day society, as many diseases are related to high air pollution. Therefore, it is important to make people aware of their contribution to air pollution and how they can diminish and deal with the risks of a high concentration of fine dust. In regard to this problem, AiREAS created an application that gives information about highly polluted areas in the region of Eindhoven. To make the application more suitable and effective, we investigated the usability and persuasiveness of this application and how it could be improved. Furthermore, we investigated which type of feedback is best suited to making people more aware of air pollution.

The persuasiveness of the current AiREAS application was evaluated using the principles of Fogg, Cialdini, and Oinas-Kukkonen and Harjumaa. These principles are a useful tool for evaluating the persuasiveness of technology. All principles scored around 3, which means that the application is not very persuasive. It is suggested that a feature be added to the application which gives users the ability to track the behavior of other users. This can lead to some sort of competition between users, which could improve their behavior. Also, giving virtual rewards will persuade users even more, because people are sensitive to compliments.

In order to evaluate the usability of the AiREAS application, we evaluated five criteria: effectiveness, efficiency, learnability, memorability and satisfaction. We also used the SUS questionnaire for measuring the usability of the application. This showed us that the application is neither very attractive nor easy-to-use. Therefore, it could be beneficial to give the application a clearer goal.

To increase the persuasiveness of the application, we investigated whether feedback could help to raise awareness among users concerning fine dust. Results show that using direct verbal feedback has the best outcome. It showed that people are more sensitive to a clear explanation as to which behavior is more desirable compared to color feedback. Implementing this type of feedback could improve awareness of air pollution and thereby decrease the risks for users.

After evaluating the usability and persuasiveness of the AiREAS application, it was concluded there are many points of improvement needed to make the application more usable and persuasive. Furthermore, direct verbal feedback gave the most promising results in making users more aware of air pollution. For future research, it might be interesting to investigate whether people are willing to choose between different modes of transportation. For example, we could research whether participants are willing to cycle to their destination instead of using the car in order to diminish fine dust production. Also, providing a personal questionnaire in which users can choose their own route to a certain destination and give direct verbal feedback could provide insight into improving their awareness of air pollution.

The AiREAS application could be a good tool for making people aware of the risks of air pollution, however it still lacks the persuasiveness and usability needed to reach its goals. Making use of direct verbal feedback has been proven to contribute to reaching these goals.

1.22 Personal Reflections of the Authors

1.22.1 Reflection: Joyce Brouns

The reason I chose the Human in Technology learning line was because I like psychology and was told that this learning line is somewhat psychological. I was not disappointed, because it did have certain psychological factors to it. Why do people behave like they do? How can you change their behavior? I was very surprised when I found out that people are so easily influenced by others and that there are actually techniques to persuade people to do something. However, when someone would ask me what I had learned from the course and whether I found it to be an interesting course, I had to say that I still have my doubts. I think it is great that everyone who is studying at the TU in Eindhoven learns something about ethics. It is important to know that there are some ethic issues behind technology, and therefore, that not everything can be done without thinking at all about the ethical issues behind it. However, it was a pity that the course consists of 20 ECTs and that one subject of 5 ECTs was not enough. You have to take all of the courses and cannot choose between different courses within the USE learning line. If the latter were possible, you could follow different subjects in different areas.

In the first course of the Human in Technology learning line, we had some guest lectures, which sometimes made the course more interesting because of the different points of view. However, sometimes it was just confusing, and different subjects could not be linked very easily, like the course about LCD television, in which someone told us in detail how an LCD television looks inside. This is interesting for some people, but it did not have a clear relationship with the other subjects in the course.

Another limitation within the course was that we had to work with people from other faculties. It was often the case that some people were doing their Bachelor End Project while others were in their first or second year. Therefore, it was hard to make appointments with each other and divide the labor. I did like the fact that we had to do some presentations, because it gave you the opportunity to improve your presentation skills in an easy way.

The project was a great opportunity to use the learning skills in practice, and therefore, I think it is a great aspect of the course, which should be kept.

In conclusion, the course was different from the subjects I follow in my primary study (biomedical engineering) and therefore interesting, but it could still use some improvements to make it a little bit more interesting. The projects were great in the way that they enabled you to bring your learning skills into practice, but it was difficult to do it with people from other years or other faculties.

1.22.2 Reflection: Tim van den Boom

I chose the Human in Technology learning line to gain a better understanding of human-technological interaction. I believed it could help me better understand the needs of technology to further improve everyday life for everyone. This course gave me great insight into how people interact with technology, how this is influenced by social context and how typical flaws are made when designers create a technological innovation. The idea of a user-centred design is something all designers need to take into account when creating a new design. This way, humans also need to be involved in the early stages of the design. These simple but crucial guidelines are helpful when designing a new technology.

I personally learned something valuable from this USE line and I really think these subjects are important for the education of every engineer. Learning how a design influences the everyday life of a user can help me realize my technical innovation and make it more useable. I do think that this USE line could have had more depth. The things we learned during certain lectures were sometimes very logical and did need to be explained. Also, some lectures were only technical and did not explain the human behavior involved at all. I especially remember the lecture by Ingrid Heynderickx in which she explained how an LCD television works and how many grey levels a television has, although she did not explain the influence on its user.

I did really like this project. Not only did it have a subject that is quite different from the subjects of my study, it also made me use the things I had learned earlier in this course. I really liked meeting with Jean-Paul Close, because it made the goals of this application very clear and helped us with our project. The collaboration among the students was good, if, however, it was sometimes a bit hard to meet up with everyone, because we were all very busy with different projects. But we divided the tasks well and we finished this report without any stress.

1.22.3 Reflection: Marjan Hagelaars

I chose the Human in Technology learning line because it was likely the learning line most linked to my bachelor in Biomedical Engineering and it fit perfectly into my schedule. However, my perception was a bit wrong, as I learned more about the psychological background than anything else. Fortunately, psychology does interest me, mostly as to how people think and their perception of products and their surroundings. In my opinion, however, the learning line is a bit too much. After one course of 5 ECTS, I had learned enough about psychology, with the two other courses of Human in Technology merely acting as repetitions of the first course. My interest faded because of these repetitions and my motivation began to decrease. The project, on the other hand, was a lot of fun. We used the information we learned in a practical sense. My conclusion therefore is that one course with lectures and one project is sufficient; in this way, you keep students motivated. In this project, I learned about the different persuasive techniques, the usability and how all these features are important for achieving an application with a goal of changing the way people think. The USE learning line could stand to become more useful for students, however the concept is clear and USEable!

1.22.4   Reflection: Relinde van Loo

I chose the USE Human in Technology course because I find the combination of technology and human interaction very interesting. The first course was mainly informative and gave a lot of different insights as to how to approach technology from the point of view of society and not the view of a scientist. All the different aspects, such as usability and the design of a product, were addressed. In the second course, I had to design a future product with my project group and evaluate both the design and ethics of this future product. These two courses were very instructive and they made me more aware of all the different sides a person can take when evaluating a product.

This project was mainly about the persuasiveness and usability of the AiREAS application and how to improve these features. I learned a lot about the different strategies of Cialdini, Fogg and Oinas-Kukkonen and Harjumaa, which provide criteria to evaluate the persuasiveness of technology. I didn't know there were so many different aspects important to persuasive technology and I found it pretty amazing to learn and study these. Moreover, this project showed me how hard it is to create technology which is persuasive and which tries to change people's behavior. I mainly focused on research question 1, which was to evaluate the persuasiveness of the current AiREAS application. I was very excited to work on this question, because persuasiveness of technology is what I find most interesting about this project. The conclusion was that the application is not persuasive and mostly provides information; different suggestions were subsequently made as to how AiREAS can improve persuasiveness. I think it is important that AiREAS makes such improvements, because their goal is to improve awareness of air pollution, but with the current application, this goal is not achieved. Their goal serves a good cause and I think that when different improvements are carried out for their application, people will be more aware of air pollution and try to reduce it.

I am very satisfied in regard to our project group and I think we efficiently divided the workload. Our collaboration went very smoothly, which contributed to the fact that we made progress with the report every week. Even though it was sometimes hard to find a time in which we all were available, we made the best of it and sometimes just worked with whoever was available at a certain time. Because of all the holidays and the busy time schedule, it was a bit unclear sometimes as to when the meeting with our supervisor, Jaap Ham, had to take place. Because of this, we missed a lot of meetings, which was a bit unfortunate. However, this did not mean that we didn't make any progress with our report, because luckily, every group member lived up to their own set deadlines for different parts of the report.

In conclusion, I think this project brought me one step closer to understanding the relationship of humans with technology.

1.22.5 Reflection: Daniëlle Ramp

I attended the USE learning line Human in Technology, with a specialization in behavioral and social theories of human technology interaction. I chose this course because of my interest in user-centred design and psychology in general.

In the first course, the focus was on empathy, which contributed to my vision and identity as a designer. In my opinion, way too many decisions are based upon assumptions nowadays. Thanks to this course, it was shown that empathizing with the users and perceiving their behavior gives us great insights and can form a good base for setting design requirements and opportunities.

The second course focused on several psychological phenomena and the online behavior of groups. I found it very interesting to learn about psychological effects, and how small changes can have a huge influence on a process. For example, whether the responsibility is centred or distributed among many has a large impact on the behavior of the individuals in a group, and therefore the group as a whole. To me, this means that if I have to make these kinds of decisions on a systemic level, I will make them very carefully and preferably based upon psychological principles.

The third course focused on persuasive technologies. Together with a group of Biomedical Engineering students, I participated in a research project on the usability and persuasiveness of the AiREAS application. I got acquainted with several theories for evaluating the level of persuasiveness (e.g., Fogg, Oinas-Kukkonen and Harjumaa and Cialdini) and made use of the System Usability Scale, a quick, reliable method for measuring the usability of a system. By means of these evaluation methods, I got to work with hard data, whereas my user evaluations were, until now, mainly based on formative feedback. My experience was that these hard data gave me a feeling of certainty, as numbers often form solid grounds for the decisions you make or the conclusions you draw. Furthermore, hard data is very useful for visualizing results, although we didn't really focus on this. For future projects with summative feedback, I would rather take this opportunity and put effort into communicating the results through clear and attractive visuals.

Although we sent questionnaires around, I wouldn't state that the user was the central point in this research, since we didn't empathize with our users. I am unable to express how our users felt, what they thought or how they acted. As a designer with an interest in social design, I strive towards personal contact with my users, and I found it frustrating and unsatisfying to miss this contact and input now.

For future design processes, this means I see added value in hard data, but to me, this value is mainly located in the area of certainty, argumentation and visualization.

Appendix I

Results of evaluation of the persuasiveness of the AiREAS application

Cialdini	Average score
Persuasive technique	
Reciprocation	3.4
Commitment and consistency	3
Social proof	1.2
Liking	4.8
Authority	2.8
Scarcity	4.4
Fogg	Average score
Persuasive technique	
Computers as persuasive tools	
Reduction	5.4
Tunneling	1.0
Tailoring	2
Suggestion	1.4
Self-monitoring	1
Surveillance	1.8
Conditioning	1.6
Computers as persuasive media	
Cause and effect	2.8
Virtual rehearsal	3.4
Virtual rewards	1.6
Simulations in real-world contexts	4.2
Computers as persuasive social actors	
Attractiveness	6.8
Similarity	5.2
Praise	2
Reciprocity	2.4
Authority	1.6
Credibility and computers	
Trustworthiness	8.4
Expertise	8

(continued)

(continued)

Presumed credibility	7.2
Surface credibility	7.8
Reputed credibility	2.6
Earned credibility	6.4
(Near) perfection	7.4
Credibility and the world wide web	
"Real-world feel"	5.6
Easy verifiability	3.2
Fulfillment	3.8
Ease-of-use	6
Personalization	2.4
Responsiveness	2.8
Persuasion through mobility and connectivity	
Kairos	1
Convenience	1.6
Mobile simplicity	7.4
Mobile loyalty	2.2
Mobile marriage	1.4
Information quality	7
Social facilitation	1
Social comparison	1
Normative influence	1
Social learning	1
Competition	1
Cooperation	1.2
Recognition	2.6
Oinas-Kukkonen and Harjumaa	Average score
Persuasive technique	
Primary task support	
Reduction	4.6
Tunneling	1.2
Tailoring	3.4
Personalization	2.6
Self-monitoring	1.8
Simulation	3.6
Rehearsal	1.2
Dialogue support	
Praise	2.6
Rewards	1.2
Reminders	2.2
Suggestions	1.4

(continued)

(continued)

Similarity	3.2
Liking	4.6
Social role 1	
System credibility support	
Trustworthiness	8.6
Expertise	8.2
Surface credibility	7.8
Real-world feel	5.2
Authority	3.4
Third-party endorsements	2.8
Verifiability	4
Social support	
Social learning	1
Social comparison	1.6
Normative influence	1
Social facilitation	1
Cooperation	1
Competition	1.2
Recognition	2.6

Appendix II: The Outcomes of the SUS Questionnaire

	ptcp. 1	ptcp. 2	ptcp. 3	ptcp.4	ptcp. 5	Average
I think that I would like to use the AiREAS Application frequently	2	1	2	1	2	1.6
I found the AiREAS Application unnecessarily complex	3	2	1	2	1	1.8
I thought the AiREAS Application was easy to use	4	4	4	3	4	3.8
I think that I would need the support of a technical person to be able to use the AiREAS Application	2	1	1	1	1	1.2
I found the various functions in the AiREAS Application were well integrated	3	4	3	3	2	3

(continued)

(continued)

	ptcp. 1	ptcp. 2	ptcp. 3	ptcp.4	ptcp. 5	Average
I thought there was too much inconsistency in the AiREAS Application	3	3	2	3	1	2.4
I would imagine that most people would learn to use the AiREAS Application very quickly	4	4	4	4	5	4.2
I found the AiREAS Application very cumbersome to use	4	4	4	5	3	4
I felt very confident using the AiREAS Application	4	3	4	3	4	3.6
I needed to learn a lot of things before I could get going with the AiREAS Application	3	4	1	1	1	2

Appendix III: The Reduced Outcomes of the SUS Questionnaire

Sum: 24.8
  Usability AiREAS app: 62 on a scale of 0–100

(1) I think that I would like to use the AiREAS Application frequently
     Score: 0.6
(2) I found the AiREAS Application unnecessarily complex
     Score: 3.2
(3) I thought the AiREAS Application was easy to use
     Score: 2.8
(4) *I think that I would need the support of a technical person to be able to use the* AiREAS *Application*
     Score: 3.8
(5) I found the various functions in the AiREAS Application were well integrated
     Score: 2.0
(6) I thought there was too much inconsistency in the AiREAS Application
     Score: 2.6
(7) I would imagine that most people would leant to use the AiREAS Application very quickly
     Score: 3.2

(8) I found the AiREAS Application very cumbersome to use
     Score: 1.0
(9) I felt very confident using the AiREAS Application
     Score: 2.6
(10) I needed to learn a lot of things before I could get going with the AiREAS
     Application
     Score: 3.0

Appendix IV: Usability Evaluation on Five Criteria

	Daniëlle	Tim	Joyce	Relinde	Marjan	Average
Effectiveness	6	2	3	3	4	3.6
Efficiency	5	5	7	8	8	6.6
Learnability	9	3	9	5	3	5.8
Memorability	8	1	7	7	2	5
Satisfaction	2	1	2	2	5	2.4
Average score	6	2.4	5.6	5	4.4	4.68

Appendix V: The Percentage of Correct Answers in the Online Questionnaire for Persuasiveness

Question number	Direct verbal feedback (%)	Direct colour feedback (%)	No feedback (%)
1	83.87	77.14	74.07
2	81.48	78.13	64.00
3	88.46	83.33	79.17
4	76.00	30.00	13.04
S	92.00	83.33	78.26
6	92.00	75.00	73.91
7	80.00	60.71	30.43
8	92.00	57.14	52.17
9	96.00	78.57	95.65
10	92.00	96.30	82.61

0HAUB0—USE-HIT-3 Projects—The Human in Technology—Research Project Report—2014–2015

Reflection by Jean-Paul Close

Persuasion in our societal context of developing core human and natural values is a new field of intelligence. It does not just redefine the characteristics of our human tools, but also the way we communicate, interact and develop our securities. Human tools for sustainable progress are all those instruments that we create that help produce our well-being, not just those from technology, but also those of supportive institutions, insights and cultural aspects. Persuasive tools in this sense are not a commercial issue, but rather one of developing common sense, awareness and responsibility. Feedback is essential for defining what works in the human psychology of *leadership and the nature of following*, and what does not. The feedback is given by the target audience themselves through involvement and by measuring achievements.

Therefore, an App directed at producing change in mentality (value oriented) is different than the act of addressing the mainstream commercially (profit oriented) around specific fragmented needs. This differentiation was relevant when trying to determine the organization that should pay for the creation of the App and if citizens would be asked or not to pay for the service?

Answering the question about citizens wanting or having to pay for the application, we can conclude:

1. When the App is used to trigger awareness and change, it should be free of charge. It is a community leadership tool, not a commercial one. Government should invest in the development using our tax money, justified by the public value development of health (in this case).
2. Once change is accepted by the mainstream, then it is no longer a leadership issue, but one of simple management for growth. The new reality becomes part of a (new) culture in which Apps and other services can be deployed as commercial items. Then, there is no persuasion anymore, just information and interaction within an accepted (new) perception of reality. In this case business invests in the development within a business case.

Persuasion is therefore only "a market" in the field of leadership development or objectives, for instance, in education, market leadership development and the desired evolution of governance into a new level of regional development. It is not a commercial item for the masses. Once persuasion has done its work, a culture change becomes visible with adapted market mechanisms and content, which, in essence, works the same as usual in the economy of transaction. It therefore can contribute to the successful establishment of unique and authentic new markets. Persuasion is hence a leadership instrument of the Transformation Economy of change (sustainable leadership side in the drawing below), creating the "new sense of what's normal", while marketing is instrumental to management (side of financial management) within the competitive Transaction Economy of growth.

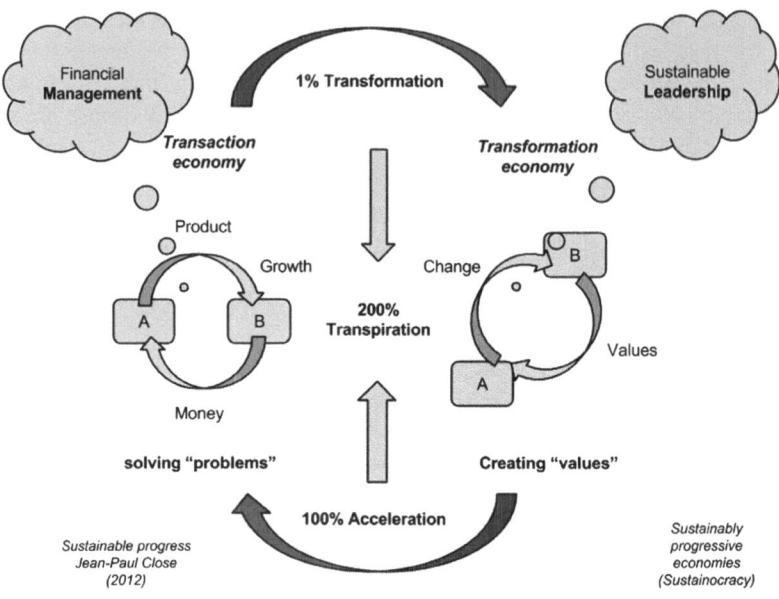

Transformation versus Transaction economies

From a regional development point of view, the new paradigm of core values, like health and air quality, positioning ourselves as peer 4 multidisciplinary co-creation, with persuasion on the side of "sustainable leadership", will entirely redefine the economy of growth (see drawing above). It is a clear reference to the way Kondratiev economic waves develop and can even be influenced positively to avoid economic crises in the future.

This particular group of students had to come to terms with this understanding and did an excellent and inspiring job. Their personal remarks are significant and part of a learning curve in which we are all engulfed. We learn as much from our pupils as they do in the process. That is why we call it *participative learning*, a technique that is not simply confirming and sending out known understandings for students to process. It connects students to real life issues in which they need to search for and process known understandings in order to come up with solutions and new insights themselves. It is not the science or knowledge itself that matters but the context and motivation in which it is implemented as an ingredient for measurable sustainable progression and creativity. Participative learning is a peer 4 (sustainocratic, awareness-driven eco-society) type of dealing with complexity at an early stage of individual human development, helping to develop personal leadership, creativity and awareness quickly for the benefit of universal and human innovation and harmony. We will deal with this in more depth in the other chapters.

Chapter 2
The AiREAS Proof of Principle—POP Relating Air Quality to Health and Lifestyle

Jean-Paul Close, Nicolette Meeder and John Schmeitz

2.1 Introduction

When we started using the ILM real time, air quality measurement system in 2013, the 'what next?' question arose amongst the AiREAS group. Our cooperative venture for creating a healthy city was not just meant to make visible the invisible but also to do something with what we see. The ILM gave us a first detailed glance into the quality of our own living environment. A multidisciplinary meeting at the end of 2013 came up with several new ideas that garnered the approval of the team. This meant that we would go ahead with multidisciplinary co-creation tables that could help us develop the proposed ideas into a project. A 'project' in AiREAS is always a complex initiative with an agreed-upon time span, an expected result and a commitment from the diversity of partners who lend it their resources, such as talent, energy, money, etc. Until that agreement is reached, it is only a positive intention, and not yet a project.

J.-P. Close (✉) · N. Meeder
STIR Foundation/AiREAS, Sustainocracy, Eindhoven, The Netherlands
e-mail: jp@stadvanmorgen.com

N. Meeder
e-mail: nicolette.meeder@stadvanmorgen.com; nicolette.meeder@gmail.com

J. Schmeitz
Schmeitz Advies, Best, Netherlands
e-mail: john@schmeitz-advies.nl

© The Author(s) 2016   31
J.-P. Close (ed.), *AiREAS: Sustainocracy for a Healthy City*,
SpringerBriefs on Case Studies of Sustainable Development,
DOI 10.1007/978-3-319-45620-1_2

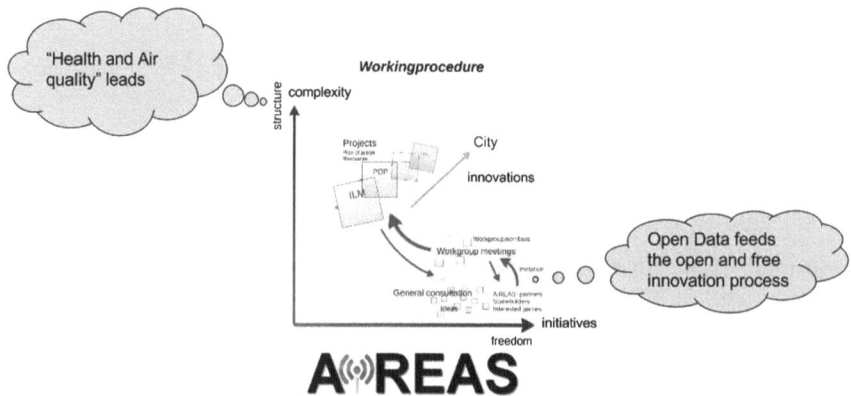

Freedom steered by core values develop innovative clusters

The effort to get from the idea stage to the project stage is coordinated by a so-called Sustainocrat, a totally independent professional who connects the many fragmented interests around the multidisciplinary, sustainocratic table. In the world of payments, the person who pays tends to lead the transaction. In the world of co-creation, money is not the leading factor, the higher purpose is. Money is a means, just like knowledge, input, technology, infrastructures, ideas, etc. Out of the co-creation, a variety of new values are obtained that can be made available again for the economy of growth. The results of our own AiREAS phase 1 had produced two show cases: the way of working at a peer 4 level and the ILM measurement infrastructure for feeding innovation through qualified real time open data gathering and interpretation. These show cases were already attracting attention from all over the world.

The POP was suggested by two different people, each with their own half of the idea. Dr. Eric de Groot is a specialist who researches heart and vascular deterioration due to air pollution. Eric suggested comparing the local exposure of people to air pollution with the effects on their health. He argued that a population of 4000 individuals would be needed to achieve results that were both scientific and practical. The individuals would represent a cross-section of the entire population and be followed over a particular period of time, for instance, 8 years. If, in the meantime, measures were to be taken in town, this should show up in the heart and vascular health evolution of this population.

Ben Nas is a social entrepreneur in one of the city quarters of Eindhoven. He is active in FRE²SH,[1] a sister organization of AiREAS within the STIR Foundation,

---

[1]FRE2SH stands for regional resilience in Food, Recreation, Energy, Education, Safety and Health through co-creation.

dedicated to developing city farming and reconnecting the city's consumption patterns with its own retriggered productivity. Ben suggested linking air quality to the reuse of space in town for primary need purposes. The rest of the AiREAS group reacted positively, suggesting that Eric and Ben's ideas might be combined. Both started to lobby for enough further support to turn the idea into a project. Eric presented the idea on numerous occasions in town, in an attempt to get sufficient civilian support. Ben started to list all of the sustainability initiatives in his city quarter to see if they could become involved or connected one way or another, rather than AiREAS reinventing the wheel and 'competing' with other pioneers. Ben came to the astonishing number of over 400 civilian initiatives in a population of 60,000, all fragmented and disconnected from each other. Each was somehow connected to the core values we were looking for. He decided to create a cycle route that connected many of the initiatives, since most did not know of each other's existence.

A view of the 'healthy city' cycle event organized by Ben and his partner Marja

With all of this human energy invested in core values, each in their own creative way, we still were not anywhere close to finding the support of the 4000 people needed to participate in the medical research. The expected financial investment for the research, spread over 8 years, was estimated to be between 12 and 15 million euros, based on similar research done elsewhere in the world. The city government was in the process of downsizing their organizational and operational costs to compensate for the effects of the credit crisis of 2008 and the growing cost of society. So, project finance needed be sourced from elsewhere (Europe or The Hague, or the other partners). Or we needed to review the project creatively to see how we might obtain the same or better results with reduced financial means. In AiREAS, we had a persistent drive to find ways to move forward, since we believe in our common purpose. We were fixated on the healthy city objective and flexible as to how we might get there. One way to reduce costs was through peer 4 multidisciplinary co-creation. People and organizations invest resources in a multidisciplinary context with their own reciprocal wishes. These are not by any means always expressed in money.

In AiREAS, we use the STIR principle of 'what exists can be bought at a cost, what does not exist can be co-created and is an investment'. Also, once a value is created, it exists and can be bought by others. New values that are created together can be inserted by the different partners into a new cycle of money-driven expansion. The interaction between value-driven motivation, the co-creation of values and the subsequent expansion of such values worldwide is referred to as the 'transformation economy', as shown in the figure previously displayed. The transformation economy is fed by data and common sense, challenging at all times the establishment with core human value-driven change. This challenge is converted into positive invitations to co-create new values when the stress at the left hand side of the drawing (the transaction economy) builds up. When the transformation economy is not taken seriously, or left to the fragmented speculation of business innovators, we see the cycle interrupted by severe crises. There are always powerful structures on the side of the transaction economy that do not want change and do everything in their power to avoid change out of self-interest. Tension builds up and can lead to an explosive situation. This can be overcome when we include Sustainocracy and the Transformation Economy structurally in the DNA of our human society.

When we were looking at the possibility of involving 4000 people in a healthy city co-creation, we invited house doctor organizations, schools, community managers, etc., to participate. No one accepted the invitation, arguing that they had no time, needed to justify themselves to their financial stakeholders, had no resources, were in crisis themselves, etc. This meant that even if we had the research budget, it would still be extremely difficult to get the right amount of participants. Also, potential partners referred to the many studies that had already been done in the world and asked what use such additional research would have?

It became clear that we needed to apply multidisciplinary persuasion to get the change going that would motivate others to join or help enlarge our mission. It could not simply be a scientific study to fill the bookshelves; it had to suit our purpose of innovative change.

2.2 E-health and Lifestyle

While looking at European funding possibilities, we found that those that might be accessible to us all had a technological and economic drive. Europe's financial backing was primarily focused on the creation of new jobs, not resolving key local issues with our core values. This had its macro-economic logic, since the hierarchy of governance is financed through taxation of consumption (VAT), contracted labor (income tax) and profitable business development (venture tax). The subsidy program in Europe hence was strongly oriented towards sustaining its own tax structure. Large multinationals were dealing with this lobby, but in the long run, neither were the issues solved, nor were jobs created, despite the enormous amounts of public money invested. Also, at the level of European governance, a new social economic resonance was needed, but this was not yet the reality of the situation. We had to deal with our value-driven reality, with all its potential, and the unjustifiable economic growth focus gripping Brussels.

E-health often came back in the calls, just like the development of handheld devices. But basic network requirements, health analyses and persuasion for change did not. We decided to take on the e-Health challenge within our own context of value-driven citizen interaction and arrived at the decision to include lifestyle analysis in our project using e-health investigation. This decision had huge consequences for the team, the approach and the way we needed to address our program. The mix of medical and social researches, applied technological tools and the many databases to be combined and interpreted first led us to want to see if we could establish a working format and team. The desired persuasive output would be fed by scientific and practical information-gathering and the complexity of value-driven interpretation and communication. The Proof of Principle (POP) was born and enthusiasm grew fast in the AiREAS community. A new scientific member was included from the e-Health perspective, Dr. Ir. Pierre Cluitmans from the University of Technology of Eindhoven. Subsequently, ICT became a hot issue bridging all the disciplines at the database level. The Sustainocrat level was also leveraged from a purely medical and social combination to a holistic approach within the contextual paradigm shift at the level of civilian participation through e-Health and awareness challenges. Now that the project had increased tremendously in terms of complexity and expert involvement, we needed to downsize the objectives to see if we could get it to work. Instead of 4000 citizens, we reduced the minimum initial need to 40.

2.3   The POP Flowchart

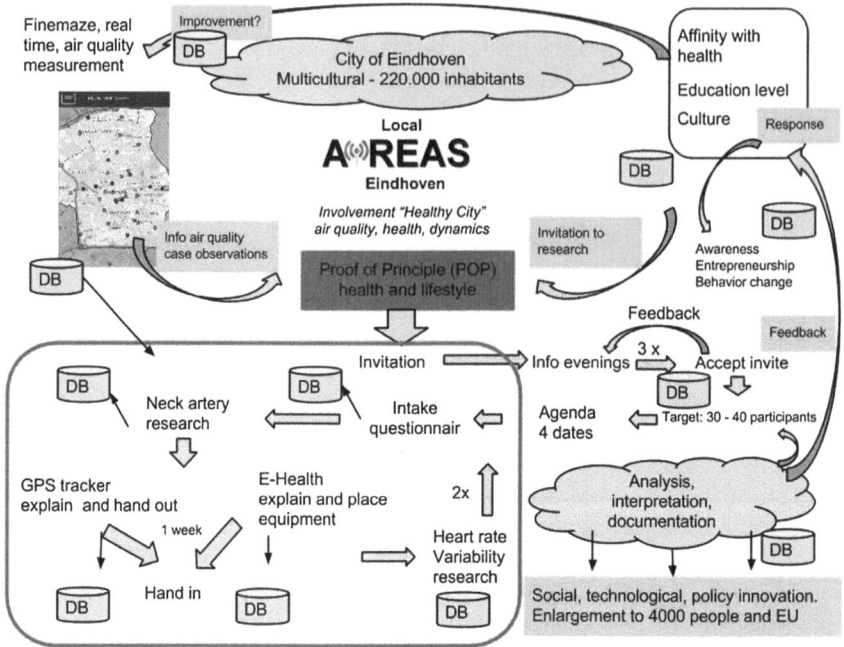

The POP flowchart

The entire process, from idea to project, with financial commitment from the local government and the multidisciplinary involvement of all required talents, had taken 8 months. It has since demanded creativity, in regard to both sticking to our higher purpose and connecting to the drivers for financial support. The team that established itself was:

Team supervisor	Jean-Paul Close
Scientifically	Eric de Groot (Vascular) and Pierre Cluitmans (HRV)
ICT	John Schmeitz, later supported by Andre van der Wiel
Civilian participation	Jean-Paul Close, Nicolette Meeder, John Schmeitz, Ben Nas
Government	Sandra van der Sterren

The first-mentioned in each category together formed the core team for discussion of progress and strategy. We engaged in dynamic clustering to deal with the specific issues whenever they arose. A book on the scientific medical and ICT insights and deliverables is written as a Phase 2 publication. This Phase 3 publication sticks to the part that deals with civilian involvement, participation and social innovation.

The POP financing was estimated to be 200,000 euros, 75 % of which was committed to by the city and 25 % by the province.

2.4   Finding 40 Participants

The enthusiasm for starting this project was high, and now that the expertise and resources had been established, one key part was still missing: *the civilian participants*. We had chosen to do the POP in a region of the city where the most pollution could be expected. Also, we determined that the local population seemed to be active when we measured the number of complaints addressed to the government referring to smells, dust on the cars and windows, etc. The local residents were united through resident's associations that defended the interests of the people in the region. When we asked the leaders of such associations of neighbors to try to gather 40 local volunteers for the research program, we were both surprised that we received so little response. We extended the call to other associations and always received the same negative attitude. Finding 40 persons was becoming an unexpected obstacle and learning curve. We had been naive in expecting people to jump at an opportunity to address an issue that had triggered their worries and corresponding complaints for such a long time.

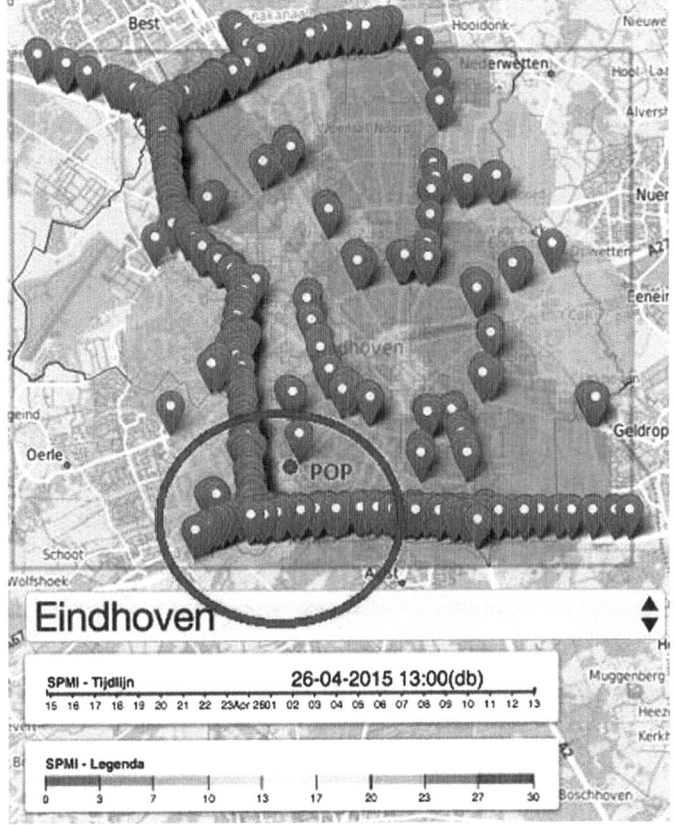

The position of the POP in town (*green areas*) related to traffic intensity (*blue*)

This taught us that people resonate with a particular socio-economic paradigm through the investment of their time and attention. This investment is related to expectations and results. Most people work all day in the current money-driven paradigm. When they come home, they want to engage in their hobbies, relax or have family time. Disturbing things like air pollution, dust accumulation, etc., are sufficient to irritate them up to the level of formulating a complaint but not necessarily to crossing the line for personal action. It is much easier to place the responsibility for the burden on the government and ask the pensioners who lead the associations of neighbors to make their complaints known, but when asked to become active in a program, other personal priorities prevail. This observation strengthened our understanding that air quality and health are leadership issues, not democratic ones for consensus, unless the socio-economic culture becomes modified. This would be the same for all sustainocratic core values.

With this in mind, one of our own participants suggested addressing the membership list of the national foundation for Environmental Defense.[2] These members already resonate with the issue of air quality and most likely would have a positive reaction to participation. The experiment was carried out and indeed a positive response was received.

We had learned that it is important to find combinations within the complex interaction of citizenship, social commitment and expectations in order to get support and participation for our program. This would be key for getting to 4000 participants. The context of people's own drive is the motivator, not just our invitation to participate. This was an important lesson, but also a worrying indicator. If we could only involve people that had the time and a context-driven motivation, we would never get to a paradigm shift that resonates primarily with such core values as health. We would only be addressing a marginal segment of the population, while the mainstream would remain uninvolved due to the reigning paradigm that produces the problems in the first place and even has a reward system backing it.

Our financial reward system stimulates people to commute between home and work and live a consumerist lifestyle. This reigning paradigm therefore rewards polluting behavior within a continuous push for economic growth. In order to transform the productive resonance of people, we would also need to address the reward system. The people who got involved in the POP did so out of their own very special motivation, which had already been confirmed by their association with and commitment to other groups. Rather than approaching it as a new paradigm, we could use the POP participation to learn about motivation, awareness, communication techniques and mentality change, even if the group had already made steps towards committing to change and the reciprocity[3] they expected. This required us to see the participants as partners in the process, rather than study objects for scientific interpretation or political influencing.

---

[2]Milieudefensie.

[3]Reciprocity became a key factor in our approach, as it refers, for us, to a much broader diversity of return on effort than the financial reward system. https://en.wiktionary.org/wiki/reciprocity.

2.5 The First Encounter

We invited the first group of confirmed participants to an evening during which we would explain what we were going to do. The evening was organized in the temporary research center that we had set up. It gave the participants the chance to see where the interaction would happen and what equipment would be used. The evening was socially oriented, with presentations by the research scientists, and drinks afterwards.

Since the participants had been defined as partners, we had decided to use the evening presentation as a living lab. After the session, the participants were contacted by telephone for feedback. This was key to improving our interaction in our attempt to get the required attitude of proactive innovation rather than them dealing with a passive research object. They were all asked to help us further through active feedback, participation and open reflection. This was to become a highly appreciated red line throughout the project. Scientific research and persuasive communication had become instruments for provoking change and our personalized interaction with the participants would provide us with measurable progress and insights into how to improve or enhance the interaction.

2.6 Feedback Collection

The feedback from our first evening presentation was an eye opener. This research was done by Nicolette Meeder. Nicolette had become a participating member of STIR right from the kickoff date in 2009. She had been active through the FRE^{2}SH experiments in food transition and the STIR Academy of participative learning. Now, she was rapidly integrated into the AiREAS POP complexity, assuming the role of mediating analyst between the participants and the project. This proved to be a fundamentally important role and one to be taken into account for repetition of the program elsewhere or its application in much larger groups.

2.7 Nicolette Meeder Reports on Feedback from the First Evening Session

The general sense of the evening was that people:

- *attended with a positive and motivated feeling*
- *felt the need to work on air quality together*
- *were prepared to attend further meetings*
- *also wanted to participate in other AiREAS activities*
- *felt this to be a nice approach, connected with other, more social cohesions*

- *felt confusion about the evening and needed clearer communication about what was expected from them.*

Suggestions for the next meeting:

- *better indication as to how to reach the location*
- *welcoming by the organizers*
- *provision of a name badge for both organizers and presenters*
- *agenda of the evening*
- *greater structure and adherence to the schedule.*

The presentations:

- *clear for some, but too scientific and complex for most*
- *one participant requested more Dutch language material rather than the abundant technical jargon in English*
- *information needed to be more practical, more focused on what was actually going to happen*
- *a short introduction to the entire objective would be preferable*
- *a clear explanation of what was expected from the participants*
- *use of fewer sheets per explanation, 1 or 2 being sufficient.*

People also liked the following ideas:

- *having room for questions and discussion*
- *having people give their names when speaking, in order to move beyond the sphere of anonymity*
- *allowing demo materials to be circulated so they might be felt, seen, touched, etc.*
- *seeing more cohesion between the speakers instead of fragmented bursts of knowledge and expertise*
- *getting the sense of a good and well-harmonized organization, giving a feeling of comfort and safety.*

In general, the participants want to be taken by the hand in a positive manner and guided through the entire process, both in terms of the structure throughout the evening but also of a more concrete plan to achieve a healthy city. The personal interviews were taken as a human and necessary experience. As a result of these, one feels seen and heard, and taken seriously in a positive progression.

2.8   End of the Report

With this, Nicolette consolidated the team through an unexpected and unplanned contribution. It confirmed for the first time the complexity of the POP and the flexibility needed to shape our progress without losing sight of our higher purpose. This would become the attitude throughout the project and key to its success.

As a consequence, Nicolette was, of course, asked to organize the second evening session with the next group of participants, using the feedback from the first. She also had to convince the diversity of research partners to adjust their interaction as a result of the feedback. This became a challenge of self-reflection for all involved. The professionals initially felt criticized and attacked in their professional self-esteem. But this was a temporary initial reaction to the new approach of extended partnership and the fixation of the focus on persuasion rather than just research. Step by step, the group of researchers also started to understand their particular role and developed a new kind of empathy with our AiREAS mission. They understood that they are not only valued for their scientific expertise, which needed no confirmation anyway. They are also recognized as scientific influencers of value-driven processes geared towards a desired end result of innovative change, measurable health and air quality development. This combination made AiREAS stand out from any other research done elsewhere in the world and lent an extra dimension to the involvement of our professionals. Apart from their islands of expertise, they are being extra-valued for their empathy with the team as a whole and their contribution to the overall objective.

2.9   Setting the Tone

After just one evening's session and feedback loop, the productive, result-driven tone was set for the entire POP from a Sustainocratic perspective. We were all now acting as a value-driven group, not as a bunch of fragmented, specialized participants. We entered a learning curve that affected every member of the team with just one objective, the health ideal of AiREAS through lifestyle, air quality and our resonance with a perceived reality.

The second evening session was done in a way that was 100 % empathic with the group. The second round of feedback was also very much more in line with the group's expectations and interaction as a team. A positive vibe had been achieved.

Nicolette became responsible for creating the agenda, having personal interaction with the participants and coordinating the interviews for the purpose of obtaining background information about the lifestyles of the people involved. Every human being may be equal, but from a biological and behavioral point of view, we all differ. The interviews were meant to provide context around each individual being examined, using various forms of medical expertise. We first researched their health by measuring the aging of their heart and arteries. We also subsequently researched the variability of their heart rate during the diverse activities that each went through during their normal day. GPS tracking determined the positioning of the person in town and their possible exposure to air pollution when outdoors in the city. Questions were also asked about the way people moved about town. There is a huge difference between riding a bike, walking and commuting by car or bus. Also, the ratio between outdoor and indoor presence is relevant for the interpretation of data.

We needed to cultivate a strong sense of responsibility in the fields of ethics and privacy. We were obtaining a lot of data from each individual that could not be shared with anyone other than the person in question. For the sake of interpretation, we needed to depersonalize the data and look at it from an objective, not subjective, point of view. The group of living objects was of interest for the study of behavior as if they were molecules in a lively environment. When providing individual feedback, the objective and subjective contexts could be brought together so that the individual might develop awareness and do something with the insight.

2.10 Data Validation

This introduced a unique new element into the investigation and usage of data. Data validation is done by experts who look at possible strange deviations that do not comply with the logic of accumulated experience. These deviations can be caused by malfunctioning equipment, errors in the programming or faults in the technological communication and database management systems. This intelligence is key to making sure that we use the right data for interpretation. But this goes for all data, including interviewing and human interaction. When combining datasets, the logic of its outcome can be judged against accumulated experience. But when these experiences don't exist, there is nothing to benchmark against. The validation of our findings then needs to be verified through dialogue between all participants, observation of the real situation and registering of the unknown for further investigation or detection of recurring circumstances. A whole new field of knowledge-gathering appeared with a complexity of interpretation against rational logic, leaving room for error and further verification.

2.11 Influencing People

During the POP's medical and lifestyle examination, the participants were not only influenced by the research itself and the data they could obtain. They confirmed that they were especially influenced by the questions providing continuous feedback in their partnership within the team and their participation in the different elements of the program. It was the questions that made the people process their thoughts. The medical and area researchers showed great empathy and took their time in explaining the working method and expected results to each of the participants individually. The need to reflect personally afterwards, in order to give relevant answers to the feedback questions, got all the partners to open up to reality in a new way. Some of the awareness triggers were:

- The opportunity to observe your own heartbeat and vascular behavior in real life and in color, without pain or invasive methods. This was an eye opener for most participants. It was also necessary to further understand the aging of the artery walls and measures people could take by themselves to improve their vascular and overall health.
- The GPS tracker that showed the whereabouts of someone during the day when plotted on a map of the city. We all live our daily routines and hardly reflect on our traces until we see it in a picture.
- Wearing instruments for heart rate variability. This made people much more aware of their daily activities, their environment, their respiration and the reaction of the body. 'Will this show on the graphs?' fostered a strange sense of curiosity and self-sensitivity. People would adjust their respiration or try to control their heartbeats simply because of the awareness that everything was being registered. This was not only done to positively influence the readings but also to experiment with the different situations they encountered in daily life. In a way, they became researchers themselves.
- The questions about registering the whereabouts of the participants during the day. These revealed that most people spend the greater part of their time indoors (90–95 % of the time).
- That lifestyle is key in determining the level of exposure to pollution and all sorts of triggers that may produce physical damage or stress.

2.12 Producing Individual Reports

The POP research with this particular group continued through June 2015. The conclusion of the physical analysis program was cause for a collective celebration, during which the scientists could present their early findings. For the medical and civilian participation teams, it all really started then. The data needed to be interpreted per set and then combined to get as complete a holistic picture of every individual as possible. This is, of course, just a moment in a lifetime, but it still gives interesting feedback and insight into our current lifestyle patterns. The POP team was not only surprised at the enormous amount of data that had been generated with a relatively small population; we also found a surprising richness in information and insight due to the cross-referencing possibilities of the different data sets. We could have spent years analyzing the data in many different compositions, but that is not the purpose of AiREAS. AiREAS wants to take steps towards developing health and a healthy city. The focus was therefore on finding ways to use the data adequately and persuasively by becoming selective. Individually, each research specialist can, of course, still do in-depth research themselves with the data obtained, but from an AiREAS perspective, it's the result-driven persuasion and communication that prevails, with measurable results to be registered in terms of positive innovation and change.

2.13   The Results

The details of this research from a medical perspective can be read in the specific publication of this AiREAS Phase 2. It took the period from June 2015 to November 2015 to interpret the different data and get to a combined analysis with feedback for each of the participants. As mentioned before, the personalized report is obviously subject to privacy. Here, we reproduce the depersonalized report as we gave it to the participants, together with their very specific private results. The very specific personal results we cannot share here.

The process was finalized through yet another feedback call from Nicolette, who tried to get a feel for the degree of change that the exercise had provoked in the mentality or behavior of the participants.

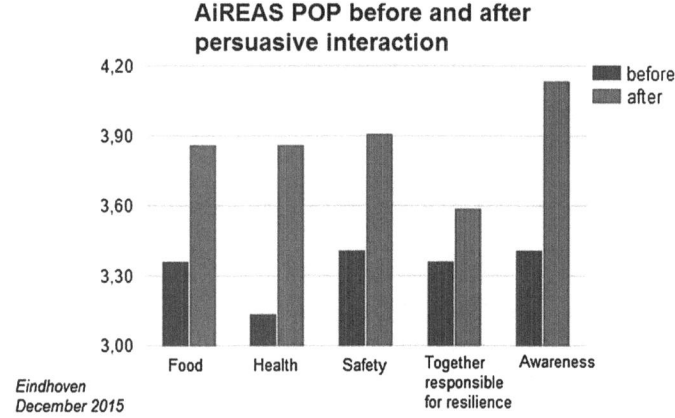

A positive evolution of awareness and change (scale 0-5)

When looking at the picture above, we need to be aware that:

- the interview took place 6 months after the physical measurements
- also, immediately after participants had obtained the personalized reports
- the group already had a high awareness level at the start (>3) due to personal choices of cohesion and participation.

Having said that, we ask ourselves what achievements we might be able to make when dealing with totally unaware and uninvolved groups?

To finalize this chapter, we share below a depersonalized copy of the final report, translated into English. The report is published under the Local AiREAS Eindhoven heading. Every AiREAS region has its own Local AiREAS platform, which allows us to allocate the innovations generated in a region to that community and share the inspiration, innovation and implementation invitation through the STIR HUB network among all Local AiREAS. Potential royalties included in the expansion process are allocated to the region where the innovation was created.

We will conclude the chapter with specific findings after the report.

Individual feedback of participating civilians in POP1

Local

Eindhoven

Individual *depersonalized* report of the results of the preliminary study on

health, lifestyle and air quality

Date: November 2015
Period(s): March 4th—11th 2015 and April 15th—22nd 2015

Local

Eindhoven

Dear participant,

You have participated in a trial project in which we tried to find a relationship between air quality and health on the one hand, and motivation, behavior, awareness and behavioral change on the other. The project, therefore, does not deal primarily with the research, but rather how it affects you as a participant and partner.

Information and communication can produce a variety of behavioral reactions, which we summarize here for you:

(1) **No reaction**: It may have been fun to participate, but it did little to nothing for me. I continue to live my life as usual.
(2) **The best defense**: I have good reasons to live as I live and will not let myself be influenced that easily.
(3) **Passive behavioral change**: I am now (more) aware of air pollution and its possible effects on my health. I will try my best to avoid the most polluted areas to reduce my exposure.
(4) **Active behavioral change**: I am (more) aware of my own possible contribution to air pollution and will try to reduce my own polluting patterns.

(5) **Innovative change**:

    (a) My consumption behavior has changed. I look very closely at what I purchase now and when something needs to be changed because it pollutes, I will look for less polluting alternatives.

    (b) I have begun to undertake action. It stimulated me so much that I became creative. I started to think of all kinds of innovations that could reduce the pollution caused by myself and others.

The AiREAS POP focuses especially on points 3, 4 and 5 with the information we supply. We are curious to see if we have managed to produce a significant contribution and will ask you for feedback at regular intervals. Of course, we will share our findings with you as well.

2.14 Introduction to AiREAS

2.14.1 What Is Aireas About?

AiREAS is a cooperative association that arose from the STIR Foundation, better known as City of Tomorrow. The STIR Foundation was founded in 2009 in response to the credit crisis and growing tension around the world as a result of the huge number of economic bubbles that have made our lives and lifestyles extremely vulnerable. STIR has been experimenting with new ways of connecting society to our core responsibilities in order to achieve sustainable (human) progress and harmonically progressive communities. From this was developed Sustainocracy, a type of democracy based on core natural and human values. AiREAS (air quality, health and regional dynamics) became one of the first sustainocratic ventures in the world. The Province of North Brabant became our Proof of Concept and the city of Eindhoven our living lab and Proof of Principle.

Along with AiREAS, other initiatives were also created. These cooperatives have the commonality that they develop core human and natural values in which citizens, local government, scientists and innovative entrepreneurs work together. Other initiatives are:

- FRE2SH and FRE2SH Farms (Food, Education, Energy, Recreation, Regional Self-Sufficiency and Health)
- STIR participative learning cooperation (Awareness, empathy and co-creation)
- STIR HUB Inspiration, innovation and implementation, sharing innovation between regions.
- SAFE, a research community examining how we can reduce our dependence on dangerous resources.

By placing your personal perception of reality at the center of our attention, combined with knowledge about air quality, mobility and health parameters, we have together made Eindhoven unique and core value-driven around health. AiREAS Eindhoven invited you to contribute to your own health as well as that of the environment in which you live and are active. Starting a local AiREAS is possible in any region, city or village.

For more information about AiREAS and a review about air quality, please consult the website: www.AiREAS.com and/or the AiREAS blog https:// globalAiREAS.wordpress.com/.

2.15   Core Values

2.15.1   Why Take Core Values as the Starting Point?

There are 5 core values, or basic principles of sustainable life, that the STIR Foundation uses to define sustainable human progress. Those core values are a part of the dynamics of all life forms, not just the human being. If a core value is not respected, it will always lead to the disappearance of life or the lifeform. We sense this as a crisis, but in reality, it is part of our evolution, and we need to adjust continuously to fit into the complexity of life in general.

We human beings can only make self-aware evolutionary decisions about our own behavior and our symbiotic relationship with our surroundings. We will deal here specifically with the human being in context with the environment of which we are an integral part based on equality. Bacteria, fungi, insects, plants, animals, fish, etc., deal with the same core values for their own evolution.

1. Food (primary resource, including drinking water and the air we breathe)
2. Health (including maintaining a healthy environment)
3. Safety (including respect for and integrity among each other)
4. Regional Self-Sufficiency (including taking responsibility together for regional mutual resilience)
5. Self-awareness (including new educational processes).

If one or more of these core values is underestimated or neglected, it will put life under pressure and our wellness will decline in all aspects. The context (how you perceive yourself, your role and life's core values) of your own situation is always the starting point. This personal context description comes primarily from the interviews that you had with Nicolette Meeder.

Our objective is that you will not just get more insight into the 5 core values, but will also become more aware of them and how they can positively influence your own situation. We hope therefore to contribute more than simple summarizing observations, by also providing you with guidance for your own future choices. For better understanding, we explore each core value here a bit more.

2.15.2   Core Value: Food

Food, water and air are the building bricks and connections of life. Food is also storage of the solar energy contained in the processes of life of the plants and animals that we eat. We eat our own evolution contained in the DNA of our system. An average human being consumes, per day:

- 1 kilo of food
- 3 kilos of water
- 30 kilos of air

What we consume is part of living and sustaining our lives, including the stored solar energy that provides us with our own energy through digestive processes. If we pollute or manipulate water, food or air, we do the same to ourselves and the billions-of-years-old mechanisms of life stored in our systems. The consequences form a genetic anthropocene[4] in which anything can happen. In exceptional cases, a new species can develop out of our own destructive processes. But mostly, the manipulating sort disappears, due to a lack of harmony with the larger context. Every human being itself is already a living harmonic mechanism that is sustained through symbiotic interaction with billions of microscopic life forms. When that is disturbed, the human being will go into decline, just as we bring our surroundings into decline. Natural food, water and air hence are core values, because we owe our physical and energetic existence to it.

2.15.3   Core Value: Health

Life is, in essence, always healthy. If unhealthy, it will die or be consumed by healthier species to provide room for a new healthy life cycle. If we accept our polluting, unhealthy socio-economic system and lifestyles, we will contribute to our own self-destruction. It is that simple.

Health is hence a core value. This core value is not just about our own health, but also that of our environment.

2.15.4   Core Value: Safety

Safety is an essential value for creating society (family, community, protection). Without safety, this would be impossible, as everyone would be stuck in the individual mode for survival rather than cohesion. Safety is key to being able to deal with the fear that goes with awareness. Safety translates into mutual respect, integrity, listening to each other, and empathy with yourself and the people around you. It relates to a place to live, clothing, warmth, equality of opportunity and societal positioning. Safety is the "right to BE". Safety protects life.

---

[4]Anthropocene - the era in which our planet's climate, atmosphere and soil are experiencing the specific results of human activities. https://en.wikipedia.org/wiki/Anthropocene.

2.15.5   Core Value: Self-sufficiency—Local Mutual Resilience

Regional self-sufficiency and mutual resilience are the basis of the creation of a society. We cannot produce all our needs ourselves and need to work as a team. But if we make ourselves dependent on a (care) system, we lose our connection to our own natural reality. We enter easily into a state of greed and apathy which becomes the basis for the appearance of hierarchies of power and control. Only regional self-sufficiency keeps us alert to the changing circumstances that require our innovative drive. Hierarchies are deadly structures when they don't deal with the core values and just focus on power and control.

This core value hence also deals with the need to take responsibility for our own local resilience and the freedom to take responsibility when needed. Freedom hence is not without obligation, but rather is obtained when, as an individual and/or community, we commit to our sustainocratic core values and bring in our talent and energy to produce common sustainable progress.

2.15.6   Core Value: Self-awareness

Self-awareness is the rational, spiritual, physical and emotional connection between the above and the wellness development of our community. It applies not only to the individual, but also to our community and the entire society format. Our world history, together with the developed scientific knowledge, provides us with all we need to progress in a wise and sustainable way, with wellness in abundance without the destruction of our habitat.

This core value also deals with the educational processes of lifelong participative learning, as developed and deployed in the STIR Academy.

2.16   Observations

In this report, we provide you with the most important observations from the research.

2.16.1   Personal Situation

In this section, we share the information we received from you during the interviews.[5]

---

[5]The statements reflected here are examples. In the real report, the actual statements of the individual are reported with feedback.

General awareness:

Participant sees more aspects that influence health (some examples):

1. Bad sleep, wakes up a few times per night due to the noise of the highway.
2. Exhaust gases, sore throats, sometimes shortage of breath.
3. Disturbance from barbecues and wood stoves.

Experiences stress when having to leave home for work.

Participant feels healthy and tries to keep it that way (examples):

1. Jogging 2× per week
2. Cycles a lot

Eliminates stress through walking and cycling.
Participant uses no medications.

We have GPS-tracker and HRV measurement data from March 4–11, 2015 and April 15–22, 2015. We made echoes on March 11 and April 22.

On your location form, you have indicated (various examples of responses):

- Outdoor activity between 1 and 3 h per day
- Most important transport device: bike

A word cloud was produced for each interview. In one glance, the participant could see the emphasis placed by him or her on the different aspects of living life.

An example of a word cloud (in Dutch)
Health (gezondheid) also appears at the core here

2.16.2   Position Determination During the Measurement Period(s)

During the pre-study, you were given a GPS-tracker with which we could map your position in town and relate it to the closest airbox. Below, you will find a map overview of all the airboxes in Eindhoven.

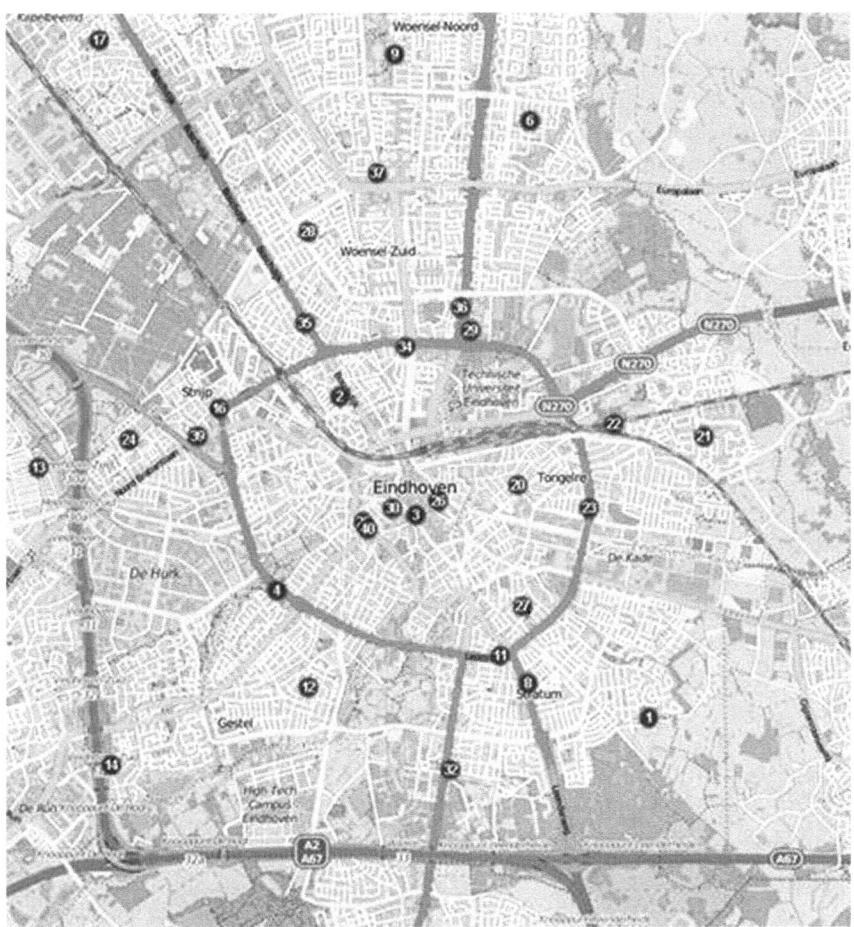

The airbox network (ILM) designed to map human exposure to air pollution
Every participant got his/her own daily GPS map

2.16.3 Exposure to Air Pollution

As we register data from the AiREAS Innovative Measurement System (ILM) in near real time, we can combine this with your position and determine your exposure to air pollution. This is just an indication, because you are not always in the direct neighborhood of the airbox, and also spend most of your time indoors. The Airbox is connected to a light post at a fixed location in town. Interpretation of your exposure is therefore always subject to modulation and indication.

The concentration of fine dust inside a building can, indeed, sometimes be up to 2.5 times higher than what we measure outdoors, depending on the situation inside and outside. There are also examples of buildings with good climate management in which the situation inside is better than that outside. Higher concentrations indoors can be caused by activities like:

- Smoking
- Open fireplace/stove
- Vacuum cleaning/changing the bed
- Cooking
- Indoor construction work
- Hobbies

For personal exposure, we have chosen the fine dust with an aerodynamic diameter smaller than 2.5 micrometers, referred to as PM2.5.

In the table below, established by the Ministry of Health, you can see that the concentration point that determines the PM2.5 transition from "Good" to "Moderate" is placed at 20 $\mu g/m^3$. We keep these reference values during our research period. If the measurement of one or more airboxes shows values above those thresholds, then we can assume a situation of the worst air quality and hence a larger possible impact on your health. The qualification of "Good" is relative too. Air pollution is always bad. Very sensitive people can also feel discomfort at lower levels of fine dust. Everything therefore depends on the particular situation of a person and environment.

The table with value indicators was established by the RIVM (Ministry of Health) in cooperation with IRAS (Risk Assessment of the University Utrecht). Information can also be found on the AiREAS site: http://www.AiREAS.com.

(µg/m³) uurwaarden	Goed			matig			onvoldoende				
	1	2	3	4	5	6	7	8			
Ozon	0-15	15-30	30-40	40-60	60-80	80-100	100-140	140-180	180-200	200–240	>240
PM10	0-10	10-20	20-30	30-45	45-60	60-75	75-100	100-125	125-150	150–200	>200
PM2,5	0-10	10-15	15-20	20-30	30-40	40-50	50-70	70-90	90-100	100–140	>140
NO2	0-10	10-20	20-30	30-45	45-60	60-75	75-100	100-125	125-150	150–200	>200
UFP*	0-2000	2000-4000	4000-6000	6000-9000	9000-12000	12000-15000	15000-20000	20000-25000	25000-30000	30000-40000	>40000
PM1*	0-7	7-10	10-14	14-20	20-27	27-34	34-48	48-61	61-68	68-95	>95

Blue good, *Yellow* moderate, *Orange* insufficient, *Red* bad, *Purple* very bad

In the graphical representation of exposure, the black line represents the threshold between 'Good' (below the line) and 'Moderate' (above the line). Underneath, in the same graphic, you find colored boxes with the numbers of the airboxes. The same color of the airbox is then traceable in the measurements. You can also find the same numbers on the back of the city map that we have shown before.

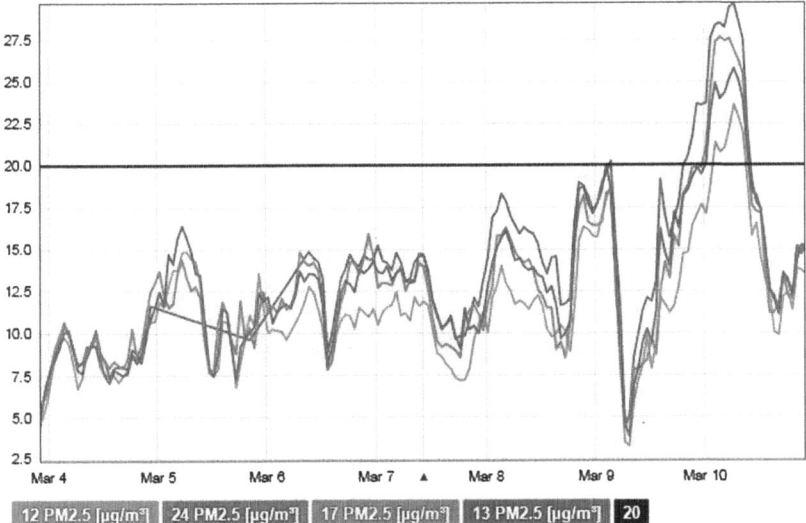

Exposure to PM2.5 of participant "x" during period "h"

There was a problem with airbox 24 on March 5, 2015, which is also shown in the graphics. Apart from that, the values remained below the norm, except for March 10. Then, the value peaked up to 30 $\mu g/m^3$.

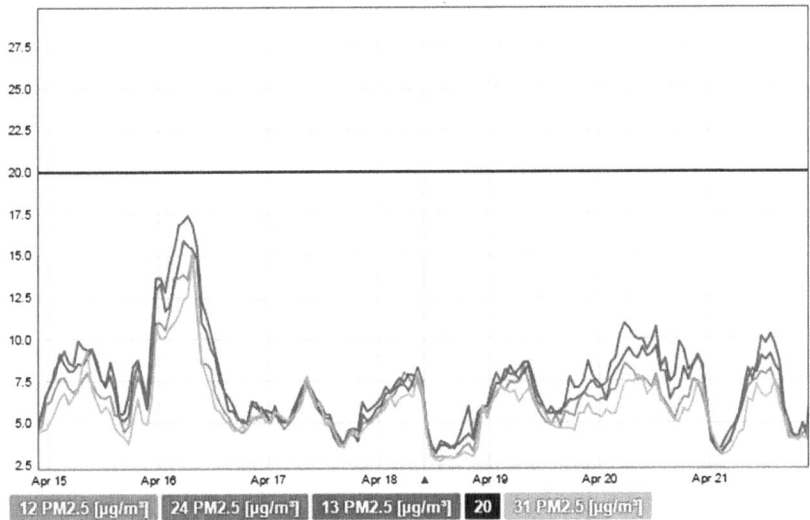

Exposure to PM2.5 of participant "y" during period "i"

Above, we see that all the values remained well under the reference line for the days of our study. Only on April 16 did the values peak a little towards the threshold.

2.16.4 Heart Rate Variability (HRV) and Motion Measurement

Twice, you were asked to wear measurement equipment on your body for one week, with which we could determine and register the variations in your heart rhythms. This indicates degrees of stress. We also register movements, allowing us to determine whether you were active or resting.

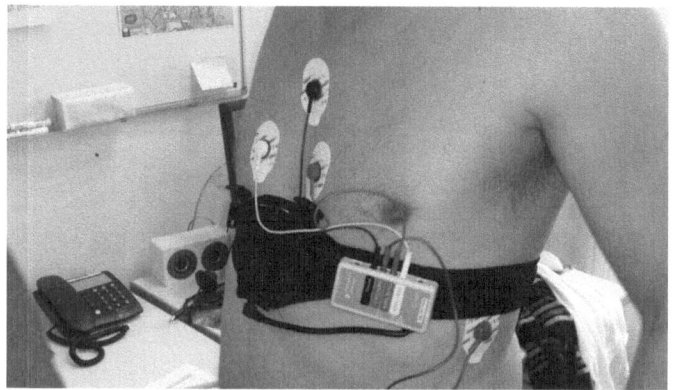

The HRV and motion sensor

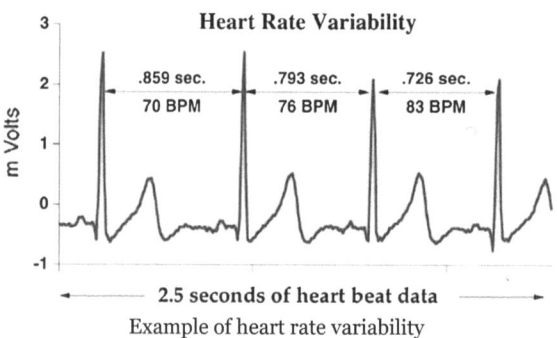

Example of heart rate variability

In the figure above, you see that the time between heart beats never remains the same. The heart is no clock. The variation is natural and is caused by our brain, which keeps regulating the rhythm. Two separate nerve tracks inside our body interact and neither can be influenced by our will. One nerve track is part of our stress system and has the tendency to make our heart beat faster. The other is part of our recovery and relax system, and has exactly the opposite effect. Both systems interact continuously in some kind of competition, sometimes with one track dominating in one heartbeat, and the other dominating in the very next. This also determines the variety we observe in the measured data.

We have accumulated a trustworthy set of data around your particular measurements. In the graphic below, we show the heartbeat intervals from beat to beat on Saturday, April 18 (participant number 3). It may look kind of messy, but our experts can still distill a reliable image of your heart rate variability during that day.

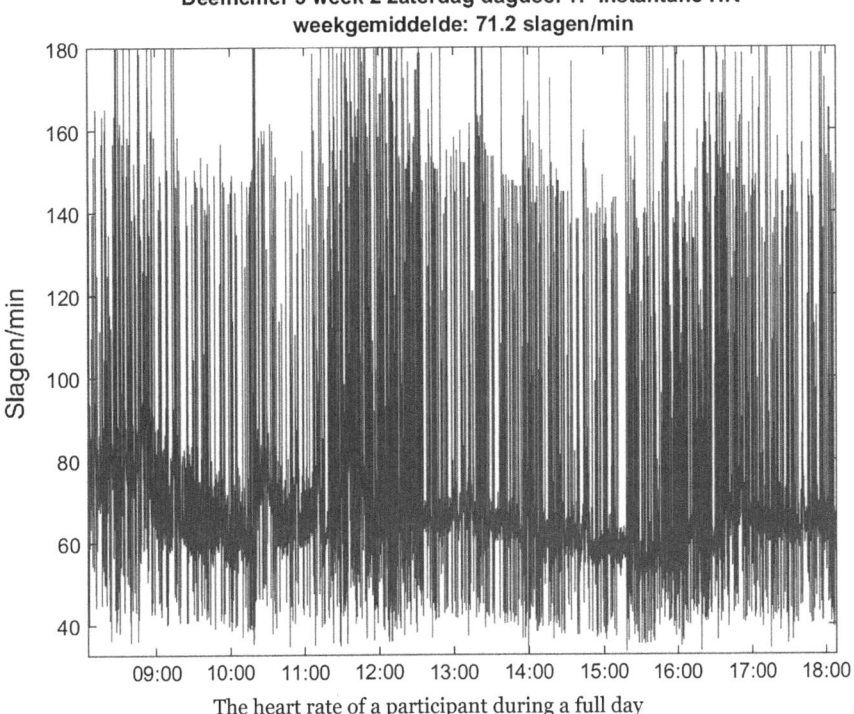

The heart rate of a participant during a full day

We can look at these fluctuations in beat to beat rhythm (HRV) in different ways. What we did with these HRV data is to relate them to the average motion velocity that we get from the GPS tracker averages and the average reference of your own for that week. No general standard values are known yet for HRV values, so this is the best we can do at this stage, and yet it already provides interesting insights.

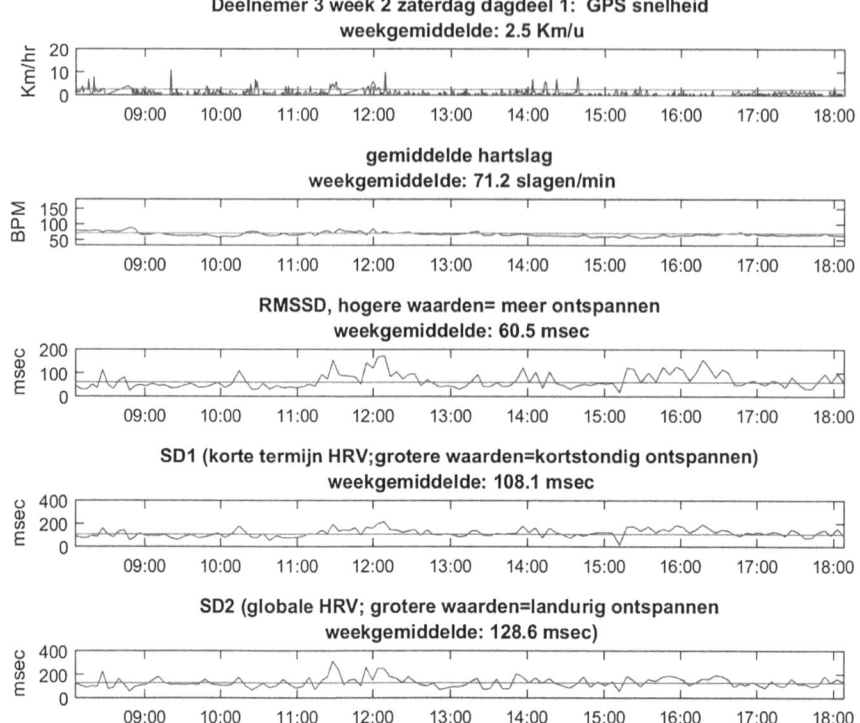

The first line in the graphic above shows the speed we determined from the GPS trackers. We can sometimes see a strange, straight line, as opposed to the mountainous curves. This means that, during that period of time, we received no data at all. Also important is the realization that if this graphic representation does not show a huge speed variation, this does not mean that an individual has not been subject to great tension, causing peaks in heartbeats. The 2nd line or part shows the average heartbeat. The other three show measurements of different HRV variables. The graphs are in Dutch. The translation of the interpretation for <u>this Saturday</u> is provided here:

km/h: Participant number and measurement week: GPS speed, weekly average of this case: 2.5 km/h
BPM: Average heartbeat. Weekly average (this case): 71.2 beats/min
ms: RMSSD, higher values = more relaxed—weekly average (this case): 60.5 ms
ms: SD1 (short term HRV: higher values = tension peak)—weekly average (just this case): 108.1 ms

ms: SD2 (global HRV: higher values = long period of tension)—weekly average (just this case): 128.6 ms

In the reference case above, personalized feedback is provided in the sense of:

The average heartbeat of just above 70 beats/min is perfectly normal. It is interesting to see that no large deviations are detected over a significant period of 10 h. In the middle graph (RMSSD), we see the most used HRV numbers. This is just one of the ways to look at the balance between the interaction of stress and relaxation. In your particular graphic, we see more fluctuations there than in the one of the heartbeat. Around noon, there is a clear moment of rest. Also, between 15:15 and 16:45, we see this period of relaxation. The same returns in the other graphs below the middle one, but is less prominently visible.

On Monday, April 20, we see another pattern:

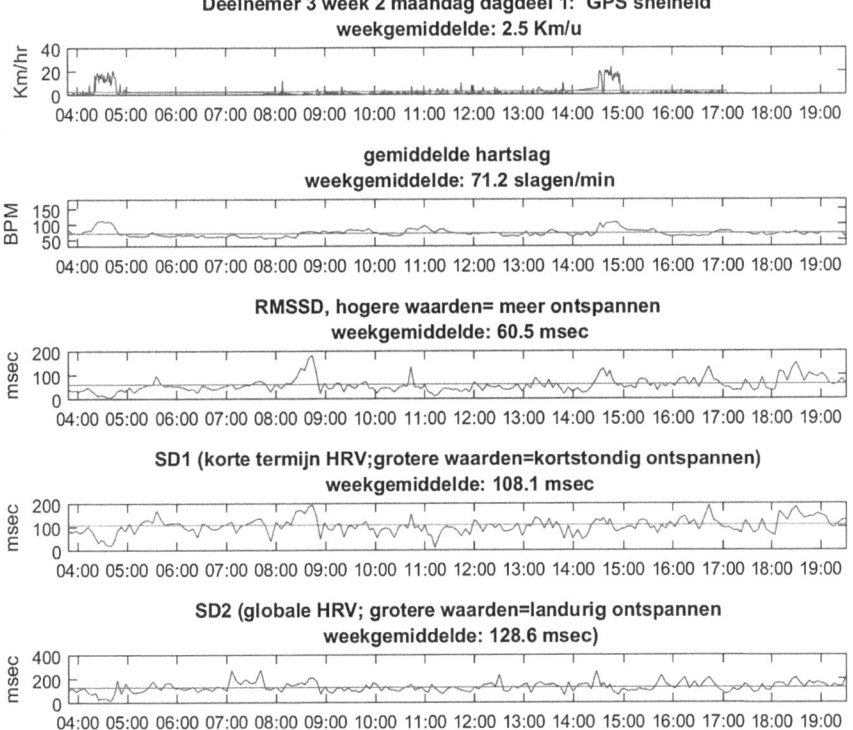

Early in the morning (between 4 and 5) and between 15:00 and 15:15, we see from the GPS speed that biking activity was most probably undertaken. The HRV graphs (RMSSD, SD1 and SD2) show a period of relaxation between 8:30 and 9:00. It is interesting is to see that, during the biking activity, relaxation seems to increase.

2.16.5 Echo Vascular Tension

To get an impression of the health of your heart and blood vessels, the left and right neck arteries (medical term: arteria carotids) were measured using modern echo techniques. In addition, the same was done with the artery in your right arm (the arteria brachialis). An echo-image is made with (inaudible) sound waves, which is non-invasive and totally friendly for the human being.

Measuring the thickness of the walls of the neck artery

The neck arteries (carotids) are important distribution channels of oxygen-rich blood to the brain. For the body, they are very large in size, with a diameter ranging between about 0.7–1.1 cm. The arteries run just underneath the surface of the skin, on the left and right sides of the neck. They relate very closely to the heart; if you put your fingers on your neck, you can feel the pulse.

Echo research of the neck artery (Dr. de Groot)

It's a fact that everyone gets older and that artery walls get thicker with age. This increase in thickness is a normal process, but is influenced by risk-factors that accelerate the thickening. Because the thickness of the arteries in the neck can easily be determined with non-invasive, painless echography, it has become one of the most popular methods for determining the long-term risk factor of heart problems. The method is relatively simple, very accurate and very well reproducible in imaging techniques. It is also a very good indicator of the condition of the rest of the blood vessels in the body.

How does this work? The thicker the artery wall gets, the greater the risk that extra local thickness will appear (atherosclerotic plaque). Sometimes, this leads to the total closure of the artery (occlusion) with all the consequences that follow. There is also an increased chance for the formation of blood clots (thrombosis) caused by this atherosclerosis. Well-known (or better yet, infamous) are the decreased bloodstream functions of the coronary arteries around the heart (coronary atherosclerosis), in the neck or elsewhere in the body, sometimes with closures and the whole host of heart and artery problems as a consequence.

From previous (echo) artery investigations over periods of more than 25 years with many thousands of participants, we know that the thickness of the carotid wall, under normal circumstances, increases with age from about 0.4 mm at birth to about 0.8 mm when you are 80 years old (blue line in the graphic below).

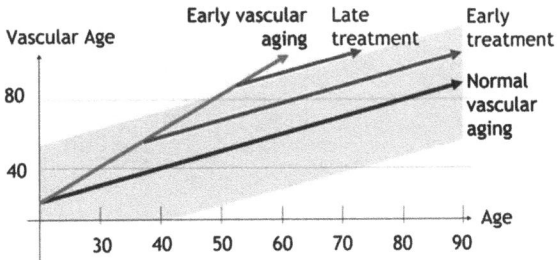

Note Exposure to risk factors such as air pollution increases the speed of vascular aging

Since getting older is something you cannot avoid, we call this a *non-modifiable* risk factor for health. The thickness of your artery wall increases (much) faster when you are exposed to particular risk factors for heart and artery illnesses (the red line in the graph). This artery thickness is an important indicator of health. In vascular medicine, a saying is used—that you are as 'old as your arteries'.

Smoking, high cholesterol, high blood pressure and exposure to air pollution are important, subject to influence *modifiable* risk factors. Nowadays, we know that if you address the modifiable risk factors, for instance, through a healthy lifestyle, the arteries get thinner and the natural thickening process once again slows down. The good news is therefore that artery thickness can be influenced in a positive sense.

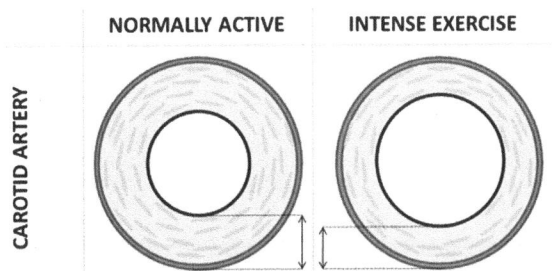

Artery wall thickness gives insight into individual health

Addressing the modifiable risk factor with healthy food, adequate physical exercise and fresh air, one can improve health before problems occur (primary prevention) or reduce the chances of renewed problems after illness (secondary prevention). The sooner one reacts, the better the possible outcome.

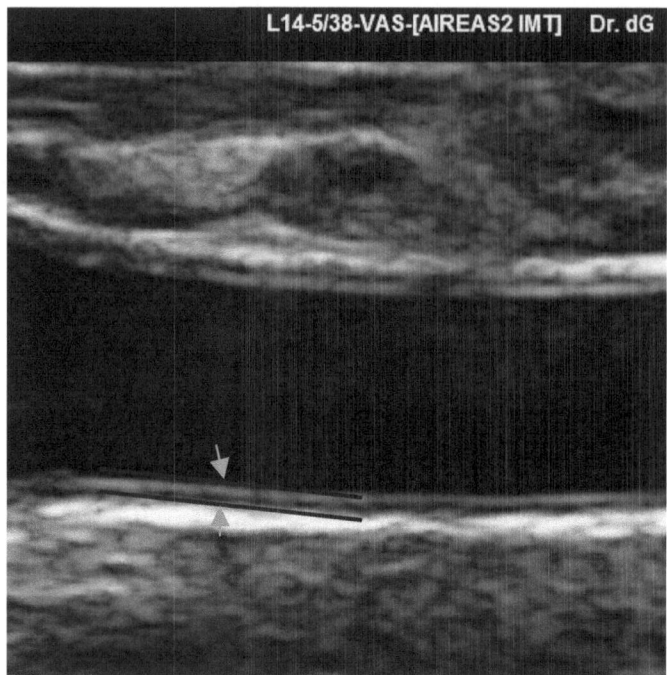

Measurement of vascular wall thickness by Dr. de Groot

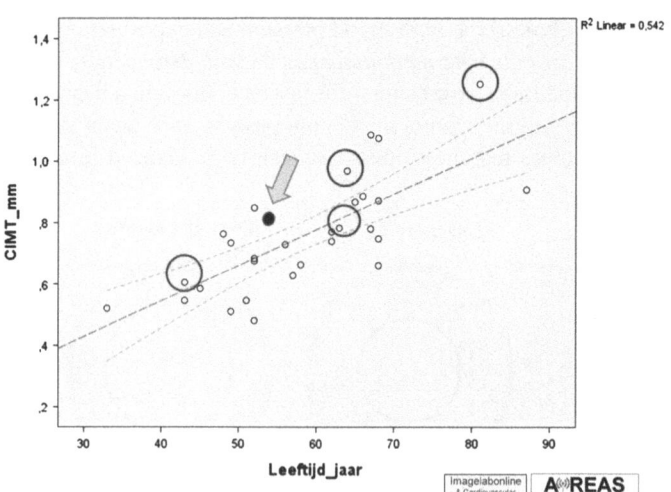

The results, mapping POP participants by age and artery thickness

Functionality of the arm artery using echo-equipment

This research was also performed by Dr. Eric de Groot and is described in detail in the book on Phase 2. The investigation concentrated on the functional vitality of the arteries. This so-called flow-mediated-dilatation (FMD) is a valid *short term* risk-indicator of the inner lining of the arteries, the *endothelial*. The wall of the artery is influenced by the (autonomous) nervous system, as well as by all kinds of molecules through receptors on the endothelial cells and in the blood stream. The endothelial, in fact, organizes the interaction between the blood stream and arterial functionality. Especially ultrafine dust (UFP) can disturb the nervous system and the endothelial function, and also therefore all vascular functionality. Long-term vascular disturbance causes an increase in the risk of heart and vascular diseases, as well as lung illnesses.

With FMD, we look at the way the blood vessel widens when the stream accelerates after applying a temporary (5 min) blood pressure cuff around the lower arm that closes the artery for a short moment.

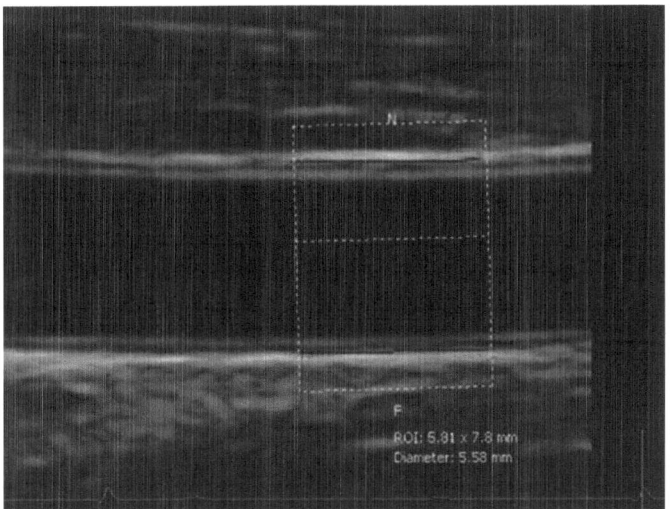

A computer image of the FMD measurement

After reducing the tension of the cuff, the bloodstream, which had built up tension due to pressure on the artery, will temporarily speed up. This increase triggers receptors on the inner lining of the artery that rapidly release a small molecule (nitrogen oxide, NO) to the muscle layer in the artery wall. This NO-release relaxes the muscular layer of the artery wall: the diameter of the artery can thus increase. This NO release and the echographically-measured blood vessel diameter increase are measures of the health of the endothelial. Since the inner lining of all the blood vessels in the body consist of endothelial, the measurements are a representative value for the artery functionality of the whole body.

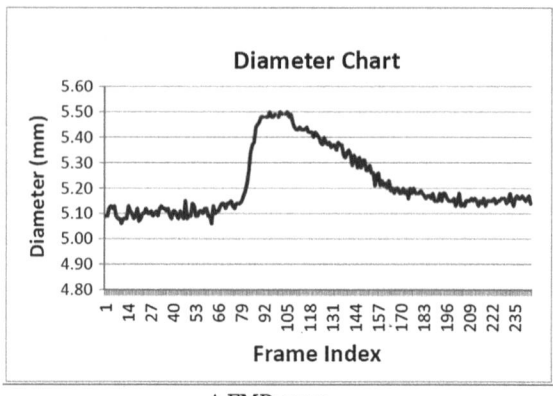

A FMD curve

A healthy endothelial shall release more NO and induce a larger widening than a less healthy endothelial. Someone with a healthy endothelial will therefore have a higher FMD.

A typical arm artery has a diameter of 3–5 mm and can widen about 5–10 % (i.e., between 0.15 and 0.5 mm!). Echographic measurement is therefore not very easy. The FMD also varies a fair amount under the influence of coffee, smoking, physical activity and food. In order to obtain a reproducible FMD, a participant needs to be sober and in a state of rest. There are no clear reference values for FMD, but the "normal" value will be somewhere between 4 and 10 %. Just as with the wall thickness, we also took average values of the available FMD measurements.

Following this information, the participant receives their personal FMD measurement and an assessment by the medical research specialist. If something really significant is detected that may prove dangerous for the participant, then the team contacts the individual so as to advise them to contact the house doctor and consult with them regarding the personalized report.

2.17 Conclusions

From all of the AiREAS POP researches, we can determine that exposure to air pollution is very much related to individual lifestyle and daily activities. The composition of air pollution also varies constantly, causing a diversity of effects on our bodies. Normally, our bodies know pretty well, through a natural drive for self-regulation based on health, how to deal with outside influences. There are, however, plenty of factors influencing the human being that directly or indirectly damage our health. Some of them we have a hand in ourselves, and with the right knowledge, we can positively influence our wellness and the quality of life for ourselves and our environment. This does not just produce better physical health, but also a much greater degree of pleasure in life, higher general productivity and an improved sense of well-being. The POP research has tried to make this visible for you.

In general, we can state that human health is determined by what we eat, breathe, drink and how we move physically with the right motivation. Awareness of these factors also alerts us to the elements that can be act as disturbances. We tend to be more aware of the origins and quality of our food. We tend to stop smoking and ventilate our houses better. We more often travel by bike or public transportation, leaving the car at home or even selling it altogether. Things that used to determine the coziness of our homes, like an open fire, candles or aromatic sticks, are replaced by other healthier, more positive alternatives. We also notice that we are talking with each other more and more often about subjects like air quality and health.

There are also things we cannot influence because they belong to our DNA-related system, or they are engrained in the reigning culture or structure of society, or they are endemic to the climate in which we live. That is why AiREAS also involves the government, science and innovative entrepreneurship in our common mission. The POP investigation is therefore valuable for influencing every aspect of society and stimulating innovation in which we can all be involved.

With our research, we try to give reassurance to everyone that this can all have a positive effect on our health and well-being, as well as a positive outcome for the global issues we face, just by giving the example and showing the results. Hopefully, you have received sufficient tips and insights to enable you to make up your own mind and take action accordingly. If you need help, you can always rely on the support of the entire AiREAS team of partners, including all specialized professionals.

In the name of the entire POP team, we thank you for your involvement and value-driven input.

2.17.1 Information Sources

The sources we used for information:

- **Exposure**: Concentrations of fine dust and Ozone during the research period via the ILM;
- **Position**: General displacement pattern during the day using a GPS tracker;
- **Speed**: Mobility based on speed via the GPS tracker;
- **HRV**: Heart rhythm variations as indicators for stress;

- **Movement sensor HRV**: Circumstances (movement or rest) in moments of stress indications;
- **Echo blood arteries**: Long-term effect of lifestyle and circumstances;

Interviews: Subject's age and perception of their health and environment.

////End of individual POP feedback////

2.18 Specific Conclusion of the POP Project

The POP was developed around two core themes: medical research and the participation of civilians. The medical research has been detailed in a specific publication (Phase 2). This current publication uses the medical outcome for measurable social innovation. We found both challenging and innovative results in developing such complexity.

2.19 Challenges Encountered

- The team member's specialties tend to work on specific islands of expertise in which each excels. The team itself and the required end results, however, need the sum of all experts, linked together through the interpretation of the different individual results in a combined complexity. In a well-integrated team, $1 + 1$ rapidly becomes much more than 2. A badly integrated team produces a negative outcome: $1 + 1 = 1.5$ or even less. We spent a lot of time becoming a well-integrated team with a positive outcome that was even greater than we expected.
- The team had to get into the habit of remembering that the participants were also part of the team and key for feedback and measurable co-creation through social innovation, executive value-driven interaction and entrepreneurship. This feedback loop proved essential for the flow, analysis and optimization of the POP.
- Our city of Eindhoven consists of many more cultures beyond just the Dutch. How do we involve all of them? We needed to experiment with this, and ended up having unexpected positive experiences with the Turkish subculture. This is described in Chap. 6.
- The HRV data contains so much detail on a microsecond level that it requires automated support to analyze such a massive amount of information. The equipment used, however, produced skin irritation for our participants. We could use the data but decided that, for a research population of 4000, we would need to find other, less irritable solutions. Only when some HRV standard exists or we determine a preferred system will we be able to develop an automated

pre-analysis of the data. The true validation and analysis is always done by human specialists and scientists through interpretation of various datasets, yet the computer can eliminate many hours of nitty gritty micro-analysis.

- We gathered 11 different databases, each with scientifically-valuable information, especially when further used in combination. This produced an immense amount of cross-referencing options for multi-data interpretation purposes. We spent a lot of time making choices based on the persuasive purpose of the POP rather than the many other interests that we could have developed around analyzing data combinations.
- If combined with other open public data sources, such as historical health information, traffic intensity, weather and climate data, the challenge risks becoming huge and chaotic without any clear added value. This may be of interest for students, but not for triggering health innovation. Choices in this field also needed to be made.
- We needed to stick very closely to the higher purpose of influencing and developing a healthy city through healthy innovation, as well as continuously managing the relationship of every research aspect to this common purpose.
- Very special and critical attention went to the ILM calibration and validation routines, since we needed to use the reliable and qualified data for our own health and lifestyle research and civilian feedback. This gave rise to the creation of a new AiREAS team, the CalVal team.
- A lot of discussion arose as to the way to report back to the participants and trigger their response. To what depth should the response go, and what might trigger awareness and even reaction? We decided always to take the positive 'can do' approach, with an invitation to take personal action for change.
- Finally, it became a challenge to get each researcher to produce publishable material. An exercise in human interaction is experienced with great professional love and passion, but the project is only finished when the documentation has been delivered. This is not just important for sharing our findings with the world; it also helps when we want to expand the activities to the 4000 people in Eindhoven or somewhere else in the world.

2.20 Innovative Results

- The POP brought us to the insight that life is always health-driven. This positive driver in a healthy city challenge produces very positive vibes among the participants.
- People tend to resonate with a reality that produces rewards. This reciprocity is not always measured in money. A learning curve, a sense of purpose, belonging to a group, secondary value development (an increase in the value of my house), worries about family members with health problems, etc., are deliverables that motivate people just as much as money may. We also introduced the AiREAS coin as a token of appreciation for value creation. With the token, people could

attend STIR evening inspiration and college sessions to learn still more about peer 4 regional development and individual participation.

- The POP got us to use technological innovations to develop research insights, but also successfully to stimulate the entrepreneurship within our own group of partners, as well as new small and medium entrepreneurial initiatives.
- The POP, combined with the ILM experiences described in our first publication, got us to redefine our interaction at the different levels of regional development and automated activities. The whole link with peer 4 (see Chap. 7) of our own City of Tomorrow views could be related to the Presencing Institute[6] and VentureSpring reports of Smart City development in Eindhoven. It came alive in the specific need for value-driven interaction between the layers. The 'large buttons' team was created to link the various levels of quality that are needed for proper interaction.

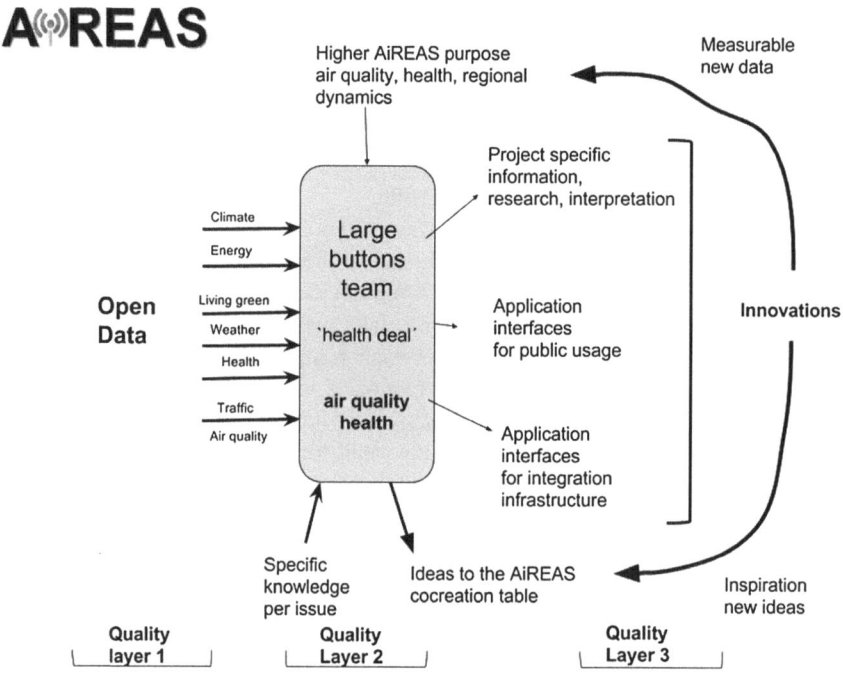

- The interaction we developed with other cultures in town, of which the Turkish community is the most representative, with a local population of about 10,000, gained us very special insights that are worth further exploration. A special

---

[6]https://www.presencing.com/ know from the U-Theory (Otto Scharmer and Peter Senge).

chapter (Chap. 6) is dedicated to this. The key insight became that core human values are creating social cohesion while political/economic drivers tend to produce fragmentation and segregation.

- We truly established a well-structured team capable of addressing this complexity in a magnificent way. We have overcome obstacles and found solutions together. The dynamics are typical of this team, due to the human bonds and cohesion that developed. It will become a challenge to train other teams elsewhere to do the same.

Despite the enormous amount of information gathered, we remained unsatisfied, because we had only covered a minute part of the city's population and the possible motivations around core values that connect or disconnect people. The medical complexity of expensive equipment and highly-trained people required us to spend a significant amount of the budget. But from a civilian participation point of view, a sufficient number of arguments had been discovered to justify trying to find new contexts and experimentation to broaden our views. POP1 had become an important part of the toolkit that a city can use to measure its health-driven progress, but we needed it to be connected to the city as a whole. We opened ourselves up to experiments, not knowing where this would take us. Being now known in the world in a new socio-economic context, with the ILM and POP1 as key instruments, we became visible to others. Our flexibility as a team did the rest as can be read in the subsequent chapters.

Chapter 3
The Backpack Project

Jean-Paul Close and Nicolette Meeder

In April 2015, AiREAS was contacted by TNO, a powerful Dutch technological research organization, who had become a member of the AiREAS value-driven community one year earlier. During that year, we had been looking at ways to get TNO involved in our projects. This time, it was TNO itself that came up with the proposition. N.B. That is how AiREAS works; anyone can take the initiative following this directive:

> If you can do it yourself within the AiREAS context of healthy city development, you do it yourself; if it's too complex, then we try to make a project and do it together.

Together with another AiREAS partner, IRAS (Health risk assessment of the University of Utrecht), TNO was looking for a place where volunteers could be equipped with measurement equipment inside a backpack for a study of their direct exposure to UFP (ultrafine particles) pollution and their lifestyle. This project, of course, fit perfectly with the objectives of the POP described in the previous chapters: the POP1 medical insights that had been obtained through the medical studies, the GPS tracker info for the public space and lifestyle questionnaires using the limitations of the fixed ILM measurement and data infrastructure. These could now be enhanced with much more direct lifestyle and exposure information, both outdoors and indoors, using backpacks. It would be difficult to get to evaluate human dynamics in a city that were much closer than this.

J.-P. Close (✉) · N. Meeder
STIR Foundation/AiREAS, Sustainocracy, Eindhoven, The Netherlands
e-mail: jp@stadvanmorgen.com

N. Meeder
e-mail: nicolette.meeder@stadvanmorgen.com; nicolette.meeder@gmail.com

© The Author(s) 2016
J.-P. Close (ed.), *AiREAS: Sustainocracy for a Healthy City*,
SpringerBriefs on Case Studies of Sustainable Development,
DOI 10.1007/978-3-319-45620-1_3

According a long-term Swedish study,[1] confirmed by our first POP steps, modern Western society spends 90 % or more of its time indoors. It is therefore very relevant for AiREAS to take indoor activities into account when defining steps towards healthy regional development. Scientific studies also indicate that, since the energy crisis in the 70s, the isolation of buildings has grown but the essence of proper ventilation has never been taken adequately into account. The indoor climate hence has suffered a great deal and, in some cases, is many times more polluted than that of the outdoors. People are not just affected by traffic, public events, construction activities and wood stoves, but also by their own cigarette smoke, perfume, indoor cleaning activities, cooking, etc. It would certainly be good to see if we could verify some of this through the 'backpack project'.

The technical aspects of this project can be read in the publication by TNO itself. Here, we concentrate on the civilian participation method, the outcome and the effects on the participants.

3.1 Getting Participation

One of the limitations of the execution of our POP project, as detailed in Chap. 2, was that we had only had access to people with a particular mindset and affiliation. This was not representative of the cross-section of an entire population. TNO had 4 backpacks available and proposed doing three sets of measurement with 4 participants, for 5 days each. The 5 days would go from Wednesday until Monday, allowing us to detect differences in behavior between weekdays and weekends. The measurements would only take place during the month of June 2015. This was a pity, because we would have liked to have gotten insight over a broader timescale, including seasonal variations and participation in all kinds of city events during the year. This was not possible, neither in practical terms nor budget-wise. We were already very happy with this window of opportunity, and AiREAS had entered the exercise without extra financial backing, so we needed to work with existing resources. The availability of space and the participants of the POP was already a great asset. Also, the experience built up during the POP, while interviewing people about lifestyle to back up the measured data, was of key importance in making the logistics, feedback collection and data analysis work, for AiREAS, but also for TNO.

The fixed setting of 4 backpacks and 5 days per session, over 3 different sessions in June 2015, spurred us into becoming creative and also introducing some diversity into the usage of the 4 backpacks. We determined three groups of people:

- 4× Participants of the original POP. This would allow us to cross-reference with the data from the POP.
- 4× Support members of City of Tomorrow or AiREAS without active involvement to date, with interesting potential contributions:

---

[1]Indoor air, the silent killer—ISBN 91-631-6161-3.

- – Reporter of the local media (newspaper)
- – Team member from the city administration (air quality specialist)
- – Residents of specific areas in town (e.g., near the airport).

- 4× TNO personnel. This would give us insight into commuting between home and work, different indoor situations (inside car, public transport, house or office …).

We were also hoping for some atmospheric or seasonal happenings in the air throughout town during the month. The groups established had the optimum mix of characteristics to obtain highly differentiated, individual lifestyle results. If an event were to occur, such as BBQ night, inversion peaks or an agricultural peak due to harvests, then we could look at the effects for the great variety of lifestyles.

An extra curiosity was the fact that TNO had recently published a very alarming report about ultrafine dust (UFP), later confirmed by the Dutch Ministry of Health (RIVM)[2] in relation to the movement of airplanes at the national airport Schiphol. This motivated us to direct some of our participants to the local Airport of Eindhoven.

One of the mixed backpack teams

Nicolette Meeder was once again in charge of the agenda, the logistics of the participants and the collection of feedback. Jean-Paul Close was in charge of the explanation of context, while TNO took care of the technicalities.

---

[2]http://www.rivm.nl/Documenten_en_publicaties/Algemeen_Actueel/Nieuwsberichten/2015/ Ultrafijnstof_door_luchtvaart_rond_Schiphol.

This time, it was easy to establish the teams. The information evening also went very smoothly. TNO came well-prepared technologically and AiREAS well-prepared in human interaction. It proved yet again a very pleasant exercise for everyone involved. IRAS, at the same time, got a car with measurement equipment to travel through town in order to get additional environmental data, but at the time of this publication, we had not received the input and hence could not take it into account in our analysis.

It took TNO over three months to analyze and interpret the data. The analysts acknowledged back that they were surprised that such a relatively small group of participants had generated such a gigantic mountain of data. Cross-referencing between questionnaires, activity logs, GPS and air quality data, all related to time, geographic position and characteristics of such location, made the interpretation very laborious and impossible to automate.

Every backpack participant received their own private report about the TNO findings. Additionally, TNO and AiREAS gave a workshop together in October 2015 with all of the participants. We wanted to share the general insights and see if people would be willing to use innovations to improve their awareness and inter-action with public spaces when making personal decisions.

3.2 The Outcome

The measurements were very surprising indeed, even though we needed to make some remarks in regard to the interpretation. For instance, when people were cooking in their homes, we would detect a peak in our data, but could not determine precisely what it was. It could have been simple water vapor, since this equipment does not distinguish materials nor does it first dry the air before counting. There is always room, therefore, for some speculation, which makes the logging of personal activities and observations very important. When someone is doing jobs in the house, it makes a lot of difference as to when the person is demolishing a wall, dusting or washing the dishes. Simply observing data does not provide the proper insight; it always needs to be cross-referenced with additional reliable information for interpretation within the proper context.

The backpack project was extensively covered by the local TV and newspaper media. One reporter had registered himself as volunteer and took part in the exercise, allowing him to report first-hand about his experiences. You can watch the TV coverage of the project here (in Dutch): https://www.youtube.com/watch?v=M7619Gzbwzw.

In general, we can reconfirm that both the contribution to air pollution and exposure is very individual and lifestyle-dependent. Lifestyle is not just the vol-untary exercise of filling the day with activities; it also has to do with the invol-untary activities imposed by labor commitments and other behavioral aspects that are defined more by the reigning socio-economic culture and paradigm than self-aware, voluntary choice. This could be easily seen in the individual readings.

3.3 The Commuters

As mentioned before, we had established three groups of volunteers. The group representing the labor-related commuters of a technology firm showed peaks corresponding to their travel time between home and work. One of those commuters lives in the rural countryside outside of town. Every morning, this person drives to work and encounters a traffic jam on the motorway. The time spent in this traffic jam shows great exposure to ultrafine dust.

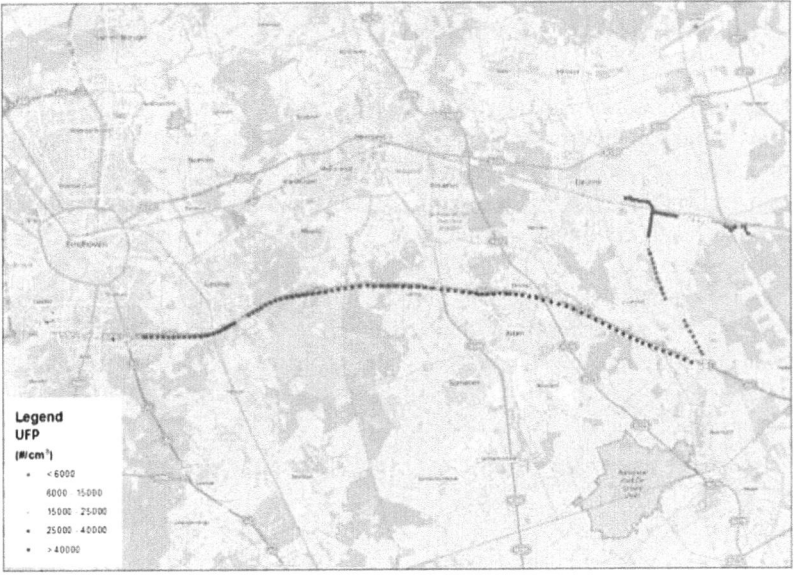

TRAVEL TO WORK DURING PEAK HOURS
(Rural to City via motoerway)

June 2015
Eindhoven region

Legend
UFP
(N/cm³)
• < 6000
   6000 - 15000
   15000 - 25000
• 25000 - 40000
• > 40000

Commuting during a traffic jam has important health consequences

Other people, those who commute by bicycle in town, find themselves exposed to occasional peaks when passing special hotspots of pollution, such as crossroads, having to stand still at a traffic light with a scooter nearby or cycling along a busy road. We see the GPS-related trail showing a diversity of ups and downs in measured values that cannot always be seen as static regions of pollution, but are merely related to the local circumstances at a specific moment in time.

Reducing the need for people to commute during peak traffic times would not only reduce personal exposure to pollution, it would also reduce the personal contribution to pollution and traffic intensity. AiREAS has suggested deploying high level internet and workspace environments in the rural villages. In our age of knowledge-based labor, people do not need to work from 9 to 5 anymore, nor be present at a specific location. Redesigning our presence at locations in time and space, using modern technological facilities close to home, can improve our wellness and well-being, creating increased productivity potential. The same can be done in the city quarters, where people can meet and contact each other in open spaces through communication facilities that network with the rest of the world.

Since all stakeholders are involved, they all start talking the same language, which, in turn, influences the local community. When the local government talks about the "health deal" as a strategic roadmap, this motivates the entire societal dialogue all the way up to and throughout the entire general population. Signs of a new socio-economic reality can be seen when looking at value-driven options relating infrastructure, individual societal contributions, lifestyle and rewards.

3.4   Lifestyle-Related Exposure

Other interesting observations in AiREAS were made while looking at the indoor activities of our participants. At home, people engage in all kinds of activities, such as cleaning, cooking, recreation, partying, smoking, etc. Lifestyle issues such as candles, stoves, individual food culture, hobbies, decoration, cigarettes, household work, furniture, etc., all contribute to pollution. This could clearly be observed in the measurement results. Again, we needed to relate the data to the logbooks of personal notes of the participants in order to find the relationships. It is clear that different seasons produce a diversity of polluting factors. In winter, wood stoves are more common than in summer, while summer features particular activities at home, such as barbecuing.

DUST PRODUCED BY DIY WEAR AND TEAR AT HOME

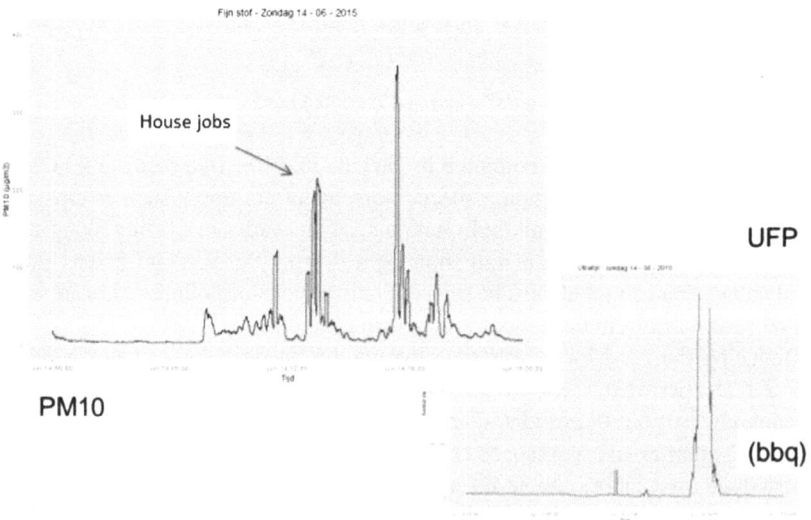

Very high peaks are related to specific activities in or around the house

In general, we observe the tendency for peaks to last longer indoors than outdoors. Outdoor dispersion of fine dust and gases is influenced very much by weather conditions, while indoor dispersion depends on ventilation.

We did not only look at high concentrations, but also observed the very low ones. Working in a clean room in a high tech environment features nearly zero exposure to fine dust. Also, the public swimming pool has a very low concentration of fine dust. Expecting a higher density of humidity in the pool, we see that we do not seem to measure water. This can be useful when interpreting household measurements.

LOCATIONS WITH REALLY LOW CONCENTRATIONS OF UFP

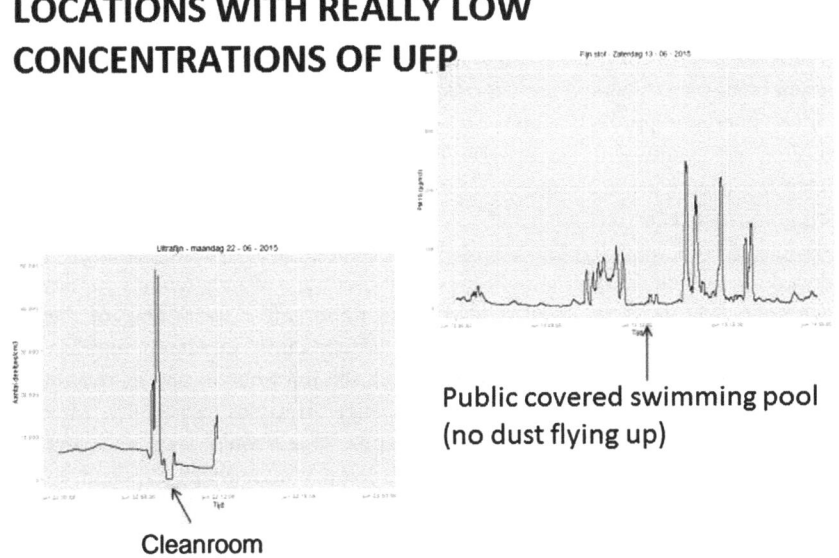

Public covered swimming pool (no dust flying up)

Cleanroom

3.5 Product Innovations

TNO presented various potential products that people could use. Examples were an App for the mobile phone, an indoor measurement system and a personal portable sensor. The participants reflected openly about the use of such devices. Many concluded that they might be useful for specific target groups, like people who already suffer from respiratory or heart problems, or those who benefit from healthy air indicators, like outdoor sport people.

People also reasoned that such services could be integrated into other standard devices that people tend to use in a home, such as a fire alarm or smoke detector. Special, personal sensor equipment was particularly seen as something one might use on those occasions when people sense the need to gain more insight. Ownership of such devices seems unnecessary and a lending mechanism could be more appropriate. Participants observe that insight does not always change behavior. One does not necessarily stay at home even when it's raining. Often, one has no choice. Apps are of interest if some interaction and personalization can be achieved.

In the end, the participants indicated that they did not want to pay much for such devices and that maybe insurance companies should get involved to co-finance investments.

3.6   Conclusions

Human beings produce and are exposed to a lot of pollution through a whole load of lifestyle-related products, habits and obligations, both indoors and outdoors. We are often not aware of it, because the pollution is invisible or not experienced as damaging. Many medical studies[3,4,5] reveal that such exposure has serious and lasting consequences for our health, as well as causing behavioral or productivity disorders, even if we still retain a sense of well-being. Much of the pollution can be reduced or avoided through social innovation, but the rest may require cultural and technological innovation, often related to an important repositioning of the way society is organized and facilitated through infrastructure, regulation and reward.

For the participants in the backpack project, the information was an eye opener, but not necessarily a major stimulus to change lifestyles, especially when the lifestyle being experienced is of high quality. One would sooner expect that innovation and regulation would resolve the main issues. Some exposure to pollution seems unavoidable, such as during one's commute between work and home. One cannot change that by oneself without challenging one's financial stability and dependence.

Contrary to the POP participants in Chap. 2, most of the people in this group were only followed up on after their personal reports had been sent out to them. Feedback was particularly necessary in regard to usage of devices or information apps. They were also only confronted with data of exposure, not their personal health. We detected a slightly less proactive attitude for change here than in the previous POP group. Since the groups are small, this is just a minor indicator and possibly subjective observation that can only be enhanced when more are involved.

---

[3]Pope III, C. Arden, et al. "Lung cancer, cardiopulmonary mortality, and long-term exposure to fine particulate air pollution." *Jama* 287.9 (2002): 1132–1141.

[4]Hoek, Gerard, et al. "Association between mortality and indicators of traffic-related air pollution in the Netherlands: a cohort study." *The lancet* 360.9341 (2002): 1203–1209.

[5]Loomis, Dana, et al. "Air pollution and infant mortality in Mexico City." *Epidemiology* 10.2 (1999): 118–123.

Chapter 4
New Entrepreneurship

Jean-Paul Close and John Schmeitz

In our theory, we expected to be able to trigger entrepreneurship while helping people to resonate with healthy city development by offering raw and processed data to which they could relate. This theory was based on our own experiences gathered through the STIR foundation while pioneering our ideology through practical entrepreneurial activities. When we moved the dot on our own horizon from competitive, career-driven to creative in the field of developing core human values, every aspect of our life's fulfilment and commitment changed. We had become entrepreneurial in a totally new sense and valued our progress through measureable steps and multiple rewards. The latter, the rewards, garnered special attention owing to the differentiation between true value and financial reward. We noticed in our own attitude that the reward of purpose-driven activity, social cohesion, meaningful exercise, a professional learning curve and social recognition was stimulating enough to overcome the burden of minimum economic reward. The economic reward structure of our society had evolved along lines that were contributing to the problems of pollution. Developing a new societal complexity also necessarily required addressing the reward system, either through the production of true sharable valuables (food, housing, energy, etc.) or through restructuring the flow of finance. A new socio-economic reality appears in which we as individuals were entrepreneurial pioneers, seeking harmonization between our purpose-driven investment, different types of reward and coverage of our daily needs.

When a whole city changes its dot on the horizon, for instance, from mainly trade- or technology-driven activity to a formal health deal with all stakeholders,

J.-P. Close
STIR Foundation/AiREAS, Sustainocracy, Eindhoven, The Netherlands
e-mail: jp@stadvanmorgen.com

J. Schmeitz (✉)
Schmeitz Advies, Best, The Netherlands
e-mail: john@schmeitz-advies.nl

J.-P. Close (ed.), *AiREAS: Sustainocracy for a Healthy City*,
SpringerBriefs on Case Studies of Sustainable Development,
DOI 10.1007/978-3-319-45620-1_4

then everything related to that city changes as well, including its choices and activities. This should also manifest itself in a new burst of vibrant, incubating entrepreneurship, developing products and services around the new paradigm. A new entrepreneurial era is introduced. Entrepreneurship in early industrial times was only profit-oriented, and processes were expressed in financial gain alone, never holding business accountable for the consequences. In 2005, I introduced this idea in my Dutch book about market leadership[1] in the 21st century with the following key differentiators:

- Before the turn of the millennium: Entrepreneurship uses humankind and the planet for financial gain
- After the turn of the millennium: Entrepreneurship serves humankind and the planet for sustainable human progress.

The entrepreneurship of the 21st century is therefore not limited to money-driven traders or speculators, but specifically connects value-driven innovators, including political executives, civil servants and those using applied scientific knowledge. Indeed, it boosts completely new structures and organizational formats for that particular kind of awareness-driven progress. Sustainocracy is a logical evolutionary step in this complexity.

We tried to prove the existence of such a boost of new entrepreneurial energy by actively looking for a positive stimulus and inviting people to do something with the challenge in an entrepreneurial way. We had already seen the initiative of civilian Ben Nas when he decided to develop a bicycle route to connect the 400 initiatives related to sustainable progress in his city quarter. This is a form of social entrepreneurship.

The problem we encountered is that the entrepreneurship that we know today needs to resonate with the financial system, not with social or ecological progress. This has led to the destruction of our surroundings and has reduced our social interaction simply because the financial system has no sentiment towards correcting itself or the moral instruments with which to do so, other than external regulation and the reaction of nature itself, including our own human ethical awareness. Regulation had created a reactive problem-solving system that was equally based on the insensitive money-driven reality. This dual economy of consumption and dealing with the effects of it has outgrown itself to critical and unsustainable proportions. The consequences were so great that the system demanded even more consumption in an economy of growth to enable government to tax, introduce insurances and more regulation to address the problems in a remedial way. When taxation does not work anymore, the national debt rises. In the Netherlands, the cost of society had increased nearly threefold in 10 years' time (2004–2014). This had an exponential character with an unrealistic prospect towards the future. Drastic changes were needed. And change means leadership with a powerful entrepreneurial spirit. This

---

[1]Handboek voor de toekomstige marktleider (2005), published by Move to Holland.

leadership was not money- but value-driven. Understanding this required a new way of looking at entrepreneurial leadership. Thus, the Pyramid Paradigm was born.

4.1   The Pyramid Paradigm

In the previous publication on the development of AiREAS and its first phase of making visible the invisible, we already went to great length in introducing the human complexity. We introduced the cyclic evolutionary progress pattern that develops between the interaction of the consequences of what we do and the discovery of what we are through backward interpretation. To DO develops our to BE. We also suggested that we are at a point in history when a major psycho-social turnaround is taking place in which we are coming to know what we are and can develop our actions and choices around that wisdom, as shown in the picture above. Only then can our awareness (to BE) start guiding our actions (to DO) (see figure below). This energetic swap is unique in human history and represents our evolution from a collective perspective. It announces a whole new era in our existence as a self-aware, creative species.

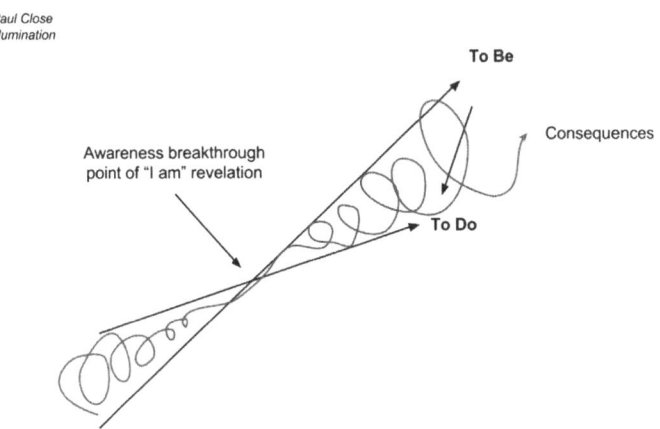

We evolve through forwards interpretation

The wisdom breakthrough and energetic swap

When we apply this to our surroundings, we see that all kinds of human beings experience this upending of their wisdom through an intense boost of awareness. When some of these people occupy leadership positions in society, they become instrumental to the overall wisdom swap of that society. When we invited people to gather for AiREAS in 2011, a large diversity of individuals attended. Some had gone through the "I am" revelation, while others had not. Those who had not were in the competitive mode of survival, looking for short-term financial gain.

They rapidly disappeared from AiREAS. The ones who had gone through this personal transformation stuck around and partnered in the multidisciplinary setting of value-driven creation and leadership. As we developed our activities, the contours of a whole new entrepreneurial setting appeared. We went step by step through this entrepreneurial breakthrough, which does not only affect business innovation but also the value-driven leadership of citizens, educators and government. We coined this as the Pyramid Paradigm.

4.1.1   The Old Money-Driven Industrial Paradigm

We are still living in the complexity of a society prior to the breakthrough of collective wisdom. This is normal, since only pioneers have crossed the line and united in preliminary value-driven settings such as AiREAS. Two worlds appear: the emerging and the one in collapse. Powerful forces try to delay the collapse, while both crises and the efforts of pioneers find openings to provide consolidated proof of the concept of the new era. This is called evolution, even though some experience it as a revolution.

The old industrial paradigm of money-driven productivity is based on a single profit-driven mechanism that consists of three basic elements: the product, the customer and cost optimization.

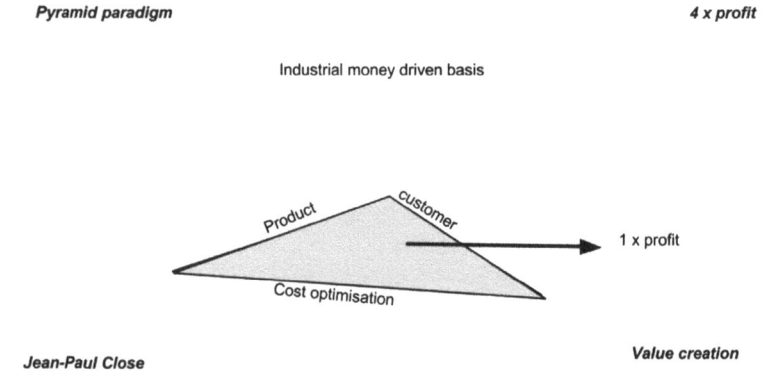

It is an established theory in business economics that when a customer can choose between 3 or more alternatives, the destruction of value appears. This is caused by the competitive drive of each of the participants in the market and the disappearance of customer loyalty due to the choice between equivalents. Cost optimization efforts can concentrate on reducing the price of the product or increasing the volume of sales. In both cases, a battle is fought which eventually will develop a shakeout among the market players at the expense of jobs, economic values and our environment. The focus on financial gain takes all emotion away from the choices made in the process, and ethics along the way. It is the law of the most aggressive that eventually wins the battle. This is, however, a temporary gain. The law of opposites shows that, while the blind battle of greed occurs, new players can introduce new ideas that eventually disrupt the greed by introducing genuine innovations. A cyclic economic pattern appears.

During the evolution of this industrial economic battle, we have seen the cyclic Kondratiev[2] patterns, showing peaks of economic development based on new communication and infrastructural innovations. The Kondratiev wave is equivalent to the cycle of human complexity of Jean-Paul Close,[3] spread out over time. This cyclic pattern is shown here:

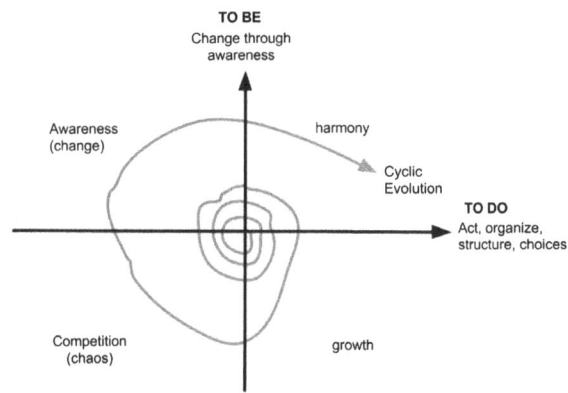

Human complexity
Jean-Paul Close (2009)

Every time an era comes to an end, a crisis develops into chaos. This develops awareness and provides openings for new innovations to deploy themselves. The innovations probably already existed but were blocked by the conservative persistence of older techniques that remained dominant. When these reached the point of market exhaustion, natural adaptation occurred, just as in the previous collapse. Humankind and economies experience this as a crisis simply because nature and human cohesion are not part of this play. The sense of a crisis is nothing other than a powerful indicator that something that used to give a sense of security has gone obsolete and something new needs to fill the gap.

Economy emulates a universe of its own, having the same mechanisms as the biological patterns of life. But owing the disconnection with the reality of those cycles of life, due to the speculative economic focus on dealing with dead things

---

[2]https://en.wikipedia.org/wiki/Kondratiev_wave.

[3]http://www.hrpub.org/download/20150620/SA5-19690341.pdf.

(matter) that human life needs to support itself, it keeps developing ups and downs without the harmony that nature always experiences, simply because it combines various value systems at once. The focus on a single currency with no other collateral than debt makes the human and ecological drama even larger. In nature, a diversity of lifeforms mingles in permanent pursuit of harmony, using different resources to develop. During a workshop at the Zoo of Emmen, the imagination of entrepreneurship was triggered through looking at the diversity of butterflies living together in a relatively small space, and peacefully at that, due to the non-competitive differentiation of size, food, reproduction, etc. Creating an ecosystem in economies can overcome problems of current models that live through single cyclic patterns.

This also has its logic in the field of human complexities. Not everyone goes through a crisis at the same time or in the same phase of their lives. Many people do this on their own and develop awareness and innovative patterns ahead of the mainstream. If their leadership is blocked by a formalized mainstream, then progress is blocked out of the system's self-interest. Negative tension then gradually builds up. However, if the leadership receives freedom to deploy itself, it generates a positive tension between the robustness of the mainstream and the argumentation of renewal. When we deployed AiREAS, we addressed the awareness level of human beings at different levels of society first. When asking a deeply aware human being about the need for core human values, hardly any resistance is felt. If this is seconded by the proven vulnerability that builds up in the institutions, then the professional position of that same human being involves making a choice: contribute to the core values through the authority of the position, or negate awareness by supporting the mission of the institution, even if it proves damaging to the core values. Awareness and guts are human factors that become decisive for taking individual entrepreneurial action, but when these factors are combined in a multidisciplinary, awareness-driven co-creation, change is a fact. The human being comes first, awareness places the core values as a permanent goal, and leadership produces the required change for harmony. We use our institutions, knowledge and technology as instruments for progress, rather than submitting to them in dependence.

We have been attempting to prove this by going through our own value-driven cycles. Every exercise in AiREAS has been developed through this method of combined entrepreneurial approach by bringing people and authorities together behind the awareness switch. A whole new dialogue appeared, including new vocabulary to express ourselves without continuous disputes about the meaning of words in the different contexts.

4.1.2   4× Profit

The new paradigm introduces the 4× profit, or Pyramid Paradigm, mechanism as an evolutionary step in entrepreneurial value creation behind the moment of the wisdom switch. Since we have now become painfully aware of the consequences of the 1× profit paradigm, the transit to 4× profit is becoming an adaptive response. Our entrepreneurial spirit does not just need financial profit through optimized processes of growth; it needs to connect emphatically to the ecological and humanitarian core values of sustainable progress through awareness and innovative change.

This evolutionary movement started around the turn of the millennium (the year 2000) with the common inspiration provided by the PPP (People, Planet, Profit) ideology. In essence, PPP introduced the other 3 profit lines of the 4× profit Pyramid Paradigm. The only confusion people experienced resulted from the different mental association around the word "Profit". For the old age's mentality, "Profit" was simply contextually related to financial gain. In this PPP societal context, it hence would still relate to the old tradition of <u>making use of the people and the planet</u> for the company's financial benefit. For the new age's mentality, "Profit" means creating measureable added value within the meaning of 'Profit = Benefit'. In this new PPP context, this would mean that financial gain (the 4th profit) would be obtained by <u>serving the people and the planet</u>. To overcome such confusion, both in entrepreneurial and in societal circles, we defined the Pyramid Paradigm within Sustainocracy. This is proof of the need for a new vocabulary representing the new energies around the new challenges, avoiding the wrong verbal and mental associations that wind up generating long, meaningless discussions rather than co-creation efforts.

Our new vocabulary of Sustainocracy has been the cause of a great deal of discussion between people in regard to their perceptions, as it triggered the curiosity of those interested in learning about its meaning. Interestingly, the use of different words and semantics already carries the sort of real energy that we represent and with which we connect to each other within ventures and projects.

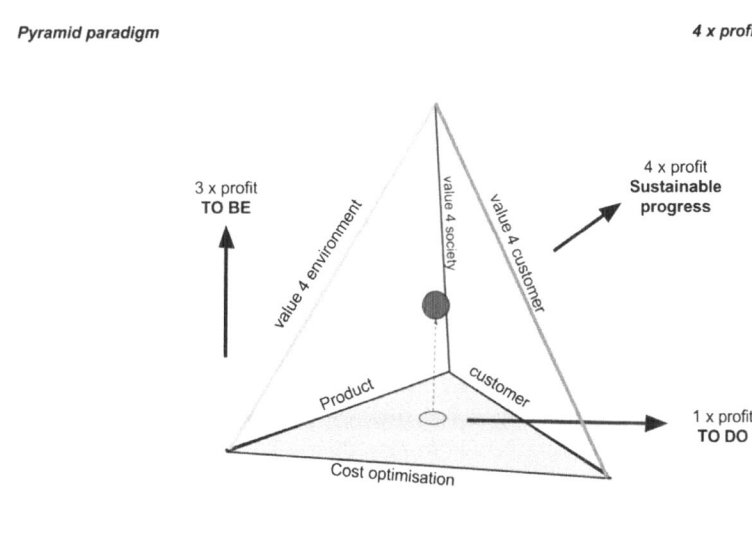

The Pyramid Paradigm put into 4 x Profit perspective

One of the consequences of such evolution is that the old, fragmented interests suddenly start to find each other in the center of that pyramid through the awareness-driven invitation. Entrepreneurship is no longer limited to money-driven business entities. Civil servants may also enter the same entrepreneurship of creating core values, not through regulation but co-creation. Civilians contribute through awareness-driven changes in their consumption patterns and productivity. The 1× profit-based business practice is outdated and evolves into value-driven co-creation, affecting every participant. A product becomes an instrument, a user too, just like the financial means, a policy or the application of knowledge. This is both a major breakthrough and a tremendous learning process for all involved.

With this, we started to experiment in order to prove the evolution of entrepreneurship of which we ourselves were an example. We now needed to show how the To Be part became dominant over our To Do decisions and that the center of the pyramid was populated with multidisciplinary tables of co-creation efforts based on core PPP + P values. Core value-driven entrepreneurship was no longer confined to "business people" but expanded so as also to include civilians, civil servants, educators, executives, etc., all of whom contributed to progress through value-driven interaction.

Entrepreneurship is no longer referred to as "making money through producing and selling"; it becomes "co-creating core values together through multidisciplinary interaction".

With this basic understanding of the evolution of regional entrepreneurship, we could start finding our way in the complex duality of the existing reality, the old field of speculative economics and the new field of economic diversity through value-driven change and awareness-driven co-creation.

4.1.3   Hackathon[4]

In Eindhoven, another partner, MAD,[5] organizes so-called Hackathons, a challenge for software developers to do something with the open data that are being generated through the town's IT internet. AiREAS was invited to participate as an open data platform with its own live stream of near real time and fine maze air quality data. John Schmeitz represented the AiREAS challenge and explained its mission. Of the 10 registered teams, 3 decided to work with the AiREAS data. One of those became the winner of this particular challenge in 2015. This shows the impact of the new entrepreneurial context presented by AiREAS. The winning team had defined a mobile application allowing people to plan their bike route through town from a health perspective, using the fine maze air quality data provided by AiREAS. The

---

[4]https://en.wikipedia.org/wiki/Hackathon.

[5]http://madlab.nl/mad/?lang=en.

idea was prize-winning, but the product did not get off the ground, because it could not connect with the economic drivers of the old paradigm. Who should finance it?

The traditional potential entrepreneurial partners concentrated on the speculative 1× profit alone. Since the App could only be deployed in Eindhoven, where we could use the ILM network, the investment would have to be covered by the local community or users. But the community is not yet in a mainstream phase of health acceptance. A commercial product was, at this stage, not feasible due to a lack of market awareness. The application would hence serve the leadership task of persuasion. It should therefore be made available free of charge and with strong persuasive techniques, as explained in Chap. 1. AiREAS has no resources of its own to finance the development other than through our partners. The local government would have been the ideal sponsor, but no one could yet convince them to (co)finance this leadership issue to be introduced free of charge into the community. The links with financial backing could not be made and the project did not materialize.

This shows yet again that money and value are two different things, and so are management and leadership, in investment patterns. To deal with this differentiation, two routes could be chosen:

1. Revolving funds can be created to support such 4× profit initiatives at the pioneer stage when the moral/ethical part is covered but the 1× economic profit still needs to prove itself through persuasion and market development.
2. A new value system can be introduced that rewards those who create value through reviewing their decisions. Think of stimulating the first people who use the app and subsequently start biking so as to reduce pollution. This has obvious short-term and long-term benefits for society, including economic. The alternative system stimulates value creation rather than trade.

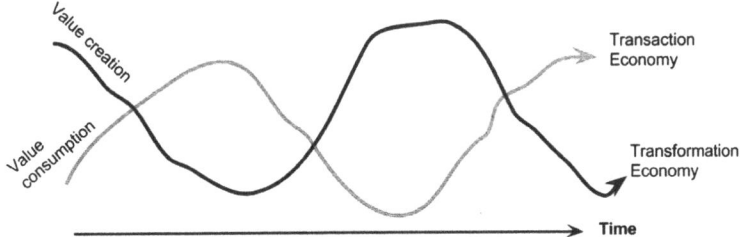

Two different value systems compensate themselves through human nature

AiREAS introduced its own AiREAS coin in an attempt to reward the value-driven participation of people. With this coin, they could access STIR inspiration classes and share locally-produced products from FRE^{2}SH. It was experimental, but did not gain immediate acceptance from the local government due to their dependence on the Euro. Meanwhile, in other regions, we saw lots of other experiments with new value systems addressing the stress created by the way in which the Euro is managed as a single currency. With such a new, value-driven unit, the transaction economy could be relieved and both could develop positively. But these mechanisms would only be understood when commonly accepted throughout a community. It is only a matter of time before such instruments become common practice. Experiments already show progress.

4.2   AiREAS Itself as Value

Meanwhile, other levels of entrepreneurship showed themselves to have significant influence. AiREAS had started off in a political and economic environment in 2011, connecting to the executive motivation of *innovation* (technology) and *civilian participation*. In 2015, new elections brought a further evolution of executive policy agreements. The new coalition adopted "health" as the main driver of technological and social innovation. This decision meant, for the first time, that the dot on the horizon had shifted from pure economic drivers to one of a core humanitarian value. Executive members of the local government were showing value-driven leadership, positioning themselves and their institutions at the core of the pyramid. This became a key source of inspiration for new age local entrepreneurship to do the same and seek coalition with its governance. AiREAS's original government partner, Mary-Ann Schreurs, became the government initiator of the local "Health Deal", a significant step forward towards a mature eco-society. This evolution can only be successful when broadly carried by the entire society and supported by the executive transition to the new era.

4.2.1   Historical Evolution

200 years ago, the very first constitution of the Netherlands was designed to mediate between the industrial and public interests, including a commitment to health in response to the effects of pollution by the enterprises. At a certain stage, the average age in the region was just 30 years, due to both diseases (pollution) and local criminality (wide gap between rich and poor). Pollution and social inequity motivated a democratic political economic reality to develop based on the duality of economic growth and dealing with the consequences of the contemporary cir-cumstances. Very soon after the introduction of the first constitutions, the political elite decided they needed to review their lawful commitment to health, because they felt they could not bear that responsibility. The variables were too large and beyond their scope of influence, other than the lawful option of regulating and introducing taxes for healthcare services. Thus, the fragmented structure we see today first appeared in which the government took on regulatory responsibility while struc-turing a remedial health care system that merely seeks to repair possible damage done. Over time, our remedial knowledge became so extraordinary that, together

with an unprecedented period of regional peace due to the introduction of social securities and diplomacy, the average life expectancy grew to over 80. This also became the main trigger for modern age financial stress and the choices that were made in the '70s and '80s to let go of the gold standard and allow unprecedented economic growth through speculation. Enormous amounts of money were needed to sustain such a caring model of the state based on money.

The number of economic bubbles produced by the combination of these decisions and the human characteristic of greed could not be foreseen until 2008, when the credit crisis opened the eyes of many to the crude reality of an unsustainable economic situation. A choice needed to be made. Manage the situation by injecting capital into the old system, hoping for recovery? Or foster leadership for change, using our awareness and entrepreneurial spirit to create something new? The local government, back in 2008, was driven by a sense of urgency, and chose to manage the situation through capital injection as an instant remediation of the problem. STIR, in 2009, chose to opt for leadership and the design of a new reality. This duality has become the basis for a new way of addressing regional development, differentiating between the already-mentioned leadership and management routines.[6] This applies to individual human beings, institutions and societal systems. It differentiates between the stress of holding on and the guts to let go.

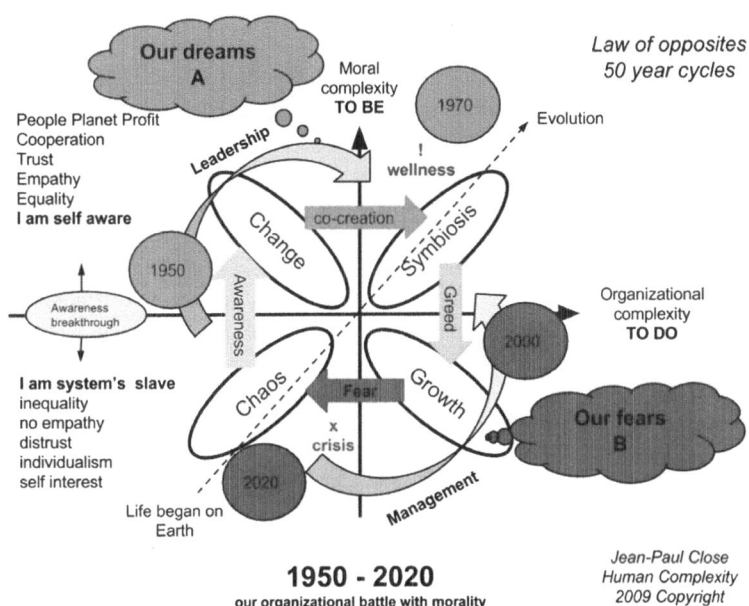

The executive dilemma: management or leadership?
"the law of opposites"

---

[6]"Succesgids voor Ondernemers", 2007—Jean-Paul Close, Pearson Education (Success guide for entrepreneurship).

The transformation of the government's financial positioning also began within that same timeframe. The investment patterns needed to change from hardware-driven infrastructural ideas to facilitating human-driven interaction, creating a transition between management and leadership through dealing wisely with chaos, letting go of fear and embracing guts.

The context of the city's directive had changed, and with it the entire image of the city's short- and long-term development. The executives started to place "healthy" in front of everything and began to create roadmaps for achieving the required results and choosing the priorities that would follow. The sustainocratic method, with a Sustainocrat as an independent chairperson, serving as a hub of connection within the multidisciplinary setting, became the recommendation carried forward for developing each of the strategic lines. This released an unprecedented amount of power, mobilizing all of the talent available in the region to contribute to the leadership trail.

What was different from 200 years ago? Why could such an evolutionary step be taken now and not back then? The answer has multiple components. The primary one is that globalization has simply reached its limits. 200 years ago, there was no global perception at all. Growth was solely the byproduct of local governance, industrial activities and labor forces, clearly differentiated pillars of society. Today, this differentiation remains in formal terms, but the practical reality has been transformed. Financial dependence, the authority of private banking systems, the system of debt and speculation of shortages, and governance of control through technology in a world market with an explosive human presence has disturbed any harmony between the people, the system and our natural environment. The citizenry has access to unlimited amounts of information that it can process for its own awareness and survival processes in an obsolete formal system. In the past, responsibility could be claimed through democratically-chosen regional governments, but nowadays, we have become aware that to change the situation, we need to let go of everything and redefine our reality, just like we did 200 years ago. The context is different, as are the issues. The perception is arising that it is not a financial issue, nor a government or business development. It is one of psycho-social awareness for which we are all responsible together, and we can only address our sustainable progress when working together through an approach of holistic regional co-creation, no longer merely thinking of the sum of the parts. The problem we must solve is overcoming the old sense of regional hierarchy that stands in the way of us finally, truly stating, 'We are all in it together'.

The only issue to solve was overcoming the idea that government itself should lead, when, in fact, the core values such as health should. Government could pave the way and behave as the backbone of the complex process, but the stakeholders had to do the innovative work together. The role of the sustainocrat had proven itself throughout the years since AiREAS had been founded and had become the connector within this complex process. Rather than the duality of "economic growth and consequences", a new duality arose: "core value-based leadership and expanding innovations". This was a far more sound economic relationship, based on the 4× Profit, and was coined the Transformation Economy, an economy focused on value-driven change rather than just growth. Harmony through change was to lead, not greed through growth.

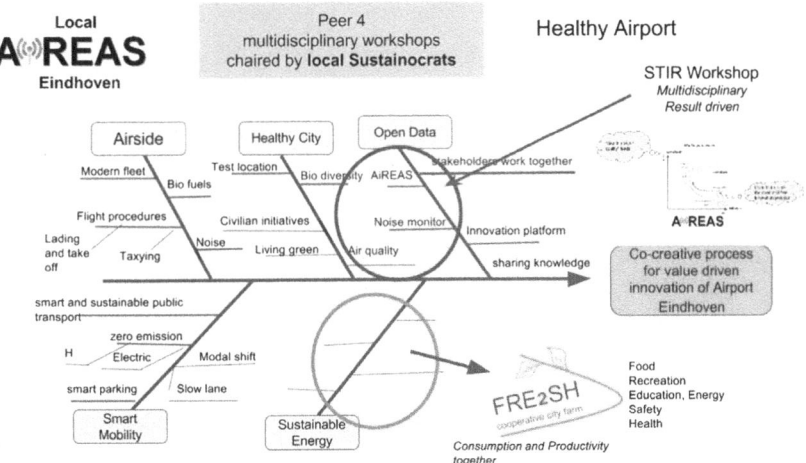

Sustainocratic roadmap for co-creating a healthy local airport

Every line along the roadmap requires value-driven entrepreneurship and intense change that can be measured along the 4× profit lines. Every creation is unique and hence a new value that can be added to the economic cycle between value creation and expansion through transactions.

4.2.2 Diner Pensant

In September 2015, the regional governance arranged an executive dinner during which the Health Deal was formulated, as well as the Sustainocratic way forward. A new era had started that had outgrown its living lab status begun in 2011. A new reality had emerged that would inspire ourselves and the rest of the world. This reality demanded a new underlying infrastructure of information and communication with which to work. It challenged the way we deal with data, leadership and each other. An initial commitment for a regional Health Deal was formulated. Eindhoven and Brabant were writing history.

The world reacted by showing interest in the format of awareness-driven multidisciplinary co-creation. In 2013, the partners within AiREAS had already determined that two core values, developed together, were ready for global expansion:

- the way of working together that had been coined Sustainocracy, including the Transformation Economy
- the phase 1 ILM structure of making visible the invisible.

4.3   Quality of Our Data and Interaction

Now that the entire regional development was resonating with the Health Deal, with the open data provided by multiple networks, including that from AiREAS, with the world watching over our shoulders, we were confronted with the imperative need to provide quality. When we established the ILM, we wanted to use the information on exposure to persuade the population to review their daily activities and develop patterns of social innovation. But the data we display and use has to be 100 % indisputable. In an experimental phase, we can still use a learning curve as an excuse, but when such basic infrastructure starts delivering data that is used for important decision-making, we need to re-examine our commitment. Not only did maintenance of the network become an issue, but also calibration, validation and interpretation. A new team was installed to examine the data independently with an eye towards three goals:

- Provide data, knowledge and feedback for policy-making
- Provide open access for users to develop their own applications and social innovations
- Connect to other systems to manage the city effectively.

The team was called the "large button" team, not because we turn those buttons ourselves, but because we influence them by guarding the quality standards of our interaction and interpretation. We established three levels of quality:

1. The origin of data
2. The capturing, CalVal and context-driven interpretation
3. The interface with the surroundings.

All this is done in a circular format, in which data produce change and this change is then captured again by the system as new data.

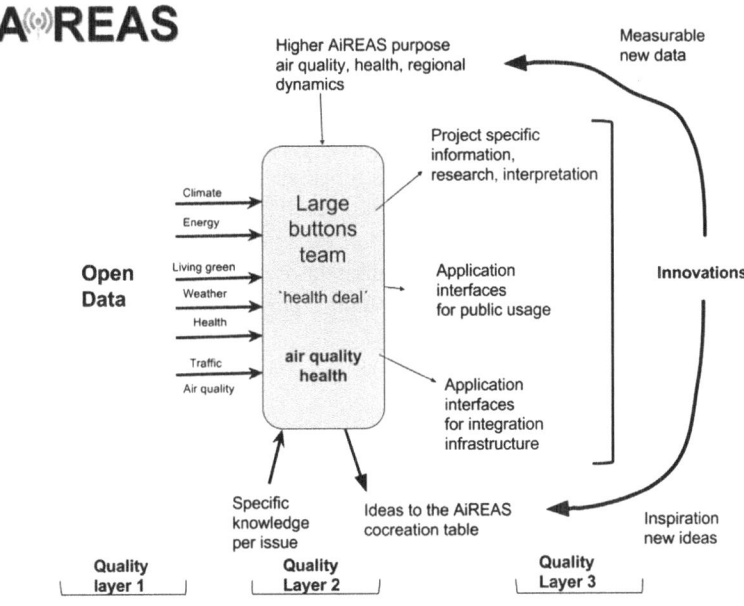

The large buttons team is, in fact, our quality assurance across the line

2015 brought about a tremendous effort to assure quality along the lines of making choices and taking steps. The original entrepreneurial partners who had produced the airboxes needed to review their commitment, since they had not taken enough into account the maintenance and calibration requirements during the operation. Low cost networks can become very expensive if they provide incorrect information, especially if the data is used to influence the entire town's dynamics. All kinds of issues needed to be resolved that had been unaccounted for in the original design and rollout of the ILM.

This learning process has been registered to avoid such issues reappearing in other regions where similar steps might be undertaken. A new economy was created that referred to the educational support for peer 4 regional development, as well as the deployment of products, services and experiences that could be used to determine network requirements elsewhere. Every region is based on the human beings that reside and live their lives there. All regions hence share the common core values but differentiate in decision-making along the lines of local priorities informed by cultural and demographical diversification. Local Sustainocrats would make the difference, as they know the complexity of the local culture and history.

4.4   Conclusion

The entrepreneurial context is relative to doing something of value for the sur-roundings in order to serve oneself. Reciprocity is another word that has a more diverse meaning than mere economic profit. It refers to the return one gets when engaging in value-driven entrepreneurial activities. The return can be to save money and resources (government), to develop or test new innovations (business), to collect new insights and knowledge (science), or to return to a healthier personal situation (civilian). All reciprocal rewards together in a multidisciplinary context do not tend to bite, but they do enhance each other. This lack of competition, with the freedom to defend one's own interests, empowers people to become entrepreneurial in a value-driven manner, no matter what talent or expectations one brings into the group, as long as they contribute to the higher purpose.

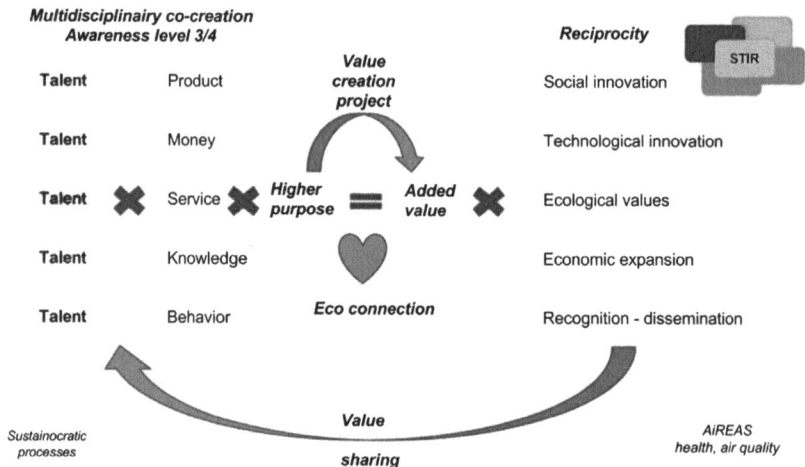

The unique interaction of a multidisciplinary table based on equality (chaired by the Sustainocrat)

A whole new world of integral, value-driven entrepreneurship developed that involved everyone within society. Reciprocity may be diverse, but a new value system is needed to compensate public creativity, particularly when it is not directly related to the speculative world of trade and euros. With such duality, stability can be assured in the region, while harmony and empathy become the leading core value-based triggers for continuous innovation. Change becomes the only constant, producing progress and a safe environment. Through the involvement of many pioneers, we started to develop such an entrepreneurial path in which the product does not lead, but rather its innovative, measurable contribution to the value-driven context does. The appearance of a product is hence conditioned to these expectations and will only be considered a success when proven through the chain of events of co-creation. This has also introduced a totally new way of rewarding the effort.

4.4.1 New Reward System

The sustainocratic table starts with no money, no budget, just the stakeholders and the higher purpose of health in relation to air pollution and regional dynamics. When a project is defined, then resources are allocated by everyone involved. Government tends to bring in public resources such as tax money and infrastructures. Others bring in innovative talent, existing or new technologies, social innovation and knowledge. So, when a project is started, the expected result is known, the commitments defined and the resources allocated. Everyone knows what to do. The money is deposited and people get paid instantly as their contribution materializes. Budgetary overshoots are not allowed and anything "unforeseen" is placed on the table as if it were a new project. The project is not finished if all commitments have not been covered and the expected results made visible. This includes the time needed to prove the effects of the multidisciplinary investment. The proof is for the benefit of all, in order that they might be able to expand the values created through contribution via the traditional transaction economy. Without this proof, the arguments that sustain the pursuit of core values are lost and the project loses at least three lines of profit. With the sustainocratic proof, the product reaches a leadership status of value creation that enhances its positioning.

4.4.2 Global Expansion

Another significant entrepreneurial activity is the expansion of our values into the global setting of evolution. The first phase of AiREAS already introduced the basics of peer 4 regional development: the unique method of working and the need for qualified information to feed innovation for the pursuit and sustainability of core values. Phases 2 and 3 will further boost this internationalization of the evolutionary steps. The immense struggle we all had to go through to structure our new reality can be largely avoided elsewhere. It also made our arguments robust and our presentation persuasive worldwide. While Eindhoven did not get the politically-desired temporary title of cultural capital of Europe in a competitive environment among cities, we had already informally claimed the title of "global capital of cultural change". Each of the pillars of our society is living up to that commitment by representing our values and co-creation. AiREAS itself, as a cooperative community, has also connected those commitments by presenting itself to the world through the following outlets:

- Open access publications, from Springer and New Horizons
- VINCI Award of Innovation
- Presentation in Barcelona, Global Expo Smart Cities
- Presentation in India: Smart Cities, Smart India
- LOI with China through Province North Brabant
- European programs H2020, Interreg V, Erasmus+
- Regional development programs with Turkey
- Healthy Airport
- Participative education program development
- Etc.

The transition affects everyone:

New entrepreneurship: sensor development, applications, drones, complex ICT data infrastructures, integrated Traffic and Air quality management structures,
New governance: Health Deal, health-based regional development, peer 4 participative society, new allocation of public funds,
New science: DAMAST, POP Health, persuasive communication, city design, etc.
Civilians: Social innovation, social entrepreneurship, participative learning.

Interestingly, we can hardly refer anymore to individual entrepreneurs, but have reached the point of an entrepreneurial society in which all participate and share the values. We still have a long way to go to make this mainstream, but the fundamentals are visible and growing, and seeds continue to find fertile ground across the world. Every year, we will see it develop itself and get more and more robust, in Eindhoven, North Brabant and throughout the whole world.

Chapter 5
Event-Linked Communication

Jean-Paul Close, Eric de Groot and Pierre Cluitmans

In August 2015, one of our partners in the regional government pointed out to us that the Marathon of Eindhoven that year was subsidized based on an innovation clause. "Would it not be nice to see if AiREAS and the Marathon organization could combine sports and air quality?" Organization of the Marathon for the coming years is in the hands of a Belgian sports organization, Golazo, which took over the responsibility from a special local NGO. The first contact with the general manager was positive and a multidisciplinary meeting was soon scheduled. Time was pressing, as the Marathon traditionally takes place in the second week of October. Golazo had already introduced innovations of their own and welcomed the partnership with AiREAS, but could not invest resources other than the available exposure and infrastructure.

As mentioned before, it is a challenge for AiREAS to reach the large civilian base of the city about the idea of them taking co-creative responsibility for their own quality of life, health and the air we breathe. Linking our objectives with the massive physical running exercise of the Marathon could give us the opportunity to see if event-related communication would be more effective than the standalone invitation to take responsibility. The context of the Marathon is much more in line with health and breathing than the daily context of our reigning social economical paradigm.

J.-P. Close (✉)
STIR Foundation/AiREAS, Sustainocracy, Eindhoven, The Netherlands
e-mail: jp@stadvanmorgen.com

E. de Groot
Imagelabonline and Cardiovascular, Eindhoven and Lunteren, The Netherlands
e-mail: ericdg@xs4all.nl

P. Cluitmans
Signal Processing Systems Group and Kempenhaeghe Expertise Center for Epilepsy, Sleep Medicine and Neurocognition, University of Technology Eindhoven, Heeze, The Netherlands
e-mail: P.J.M.Cluitmans@tue.nl

© The Author(s) 2016
J.-P. Close (ed.), *AiREAS: Sustainocracy for a Healthy City*,
SpringerBriefs on Case Studies of Sustainable Development,
DOI 10.1007/978-3-319-45620-1_5

The first multidisciplinary encounter, AiREAS - Marathon

5.1 Call for Co-creation

Within the scope of AiREAS, a call among all ILM and POP partners was made to bring a number of multidisciplinary talents into this communication channel and opportunity. "No budget" does not mean anything when dealing with creation; on the contrary, it stimulates co-creation, especially when one is surrounded with top talent. The core asset of a creative community is not money, but rather the combined talents and willingness to connect in value driven initiatives of the participants. This, in fact, is the message of the STIR Foundation, the claim that activated citizens are the main value of a city. Buildings and streets are mere instruments for producing value-driven interaction. In our perception they only represent value when in use.

The marathon challenge within the context of the POP2 of civilian participation and response to air quality was one of open experimentation and reciprocity in the diversity of possible returns. Following the AiREAS POP approach, three lines of action developed:

- Communication in an event-related environment
- Medical research among athletes and their supporters
- Value-driven entrepreneurship.

From a communication point of view, the Province of North Brabant showed a willingness to help develop materials, as long as the research and results were utilized on a provincial level, rather than merely being local event-related. From a sustainocratic point of view, this is a logical approach. The Marathon is not our goal, health and healthy air is, with the Marathon as means. This set the tone for our investigation, which also involved the cities of Helmond and Breda in the challenge. It was clear to us that Marathon runners do not just come from Eindhoven, but also from other regions. They train where they live, so involvement of a broader geographical scope beyond Eindhoven was certainly desirable. It also connected to a broader recreational philosophy of peer 4 regional development, connecting urban and rural infrastructures in such a way so as to stimulate sport, recreation and physical exercise.

In this study, we have already concluded that modern people spend about 90 % of their time indoors while sitting down. Even outdoors, we spend most of the time sitting in or on vehicles. Sitting is now being called the "new smoking", the next major health hazard that our lifestyle has developed. When we realize that we consume 30 kg of air every day, the context of sitting down indoors versus physical exercise outdoors provides interesting research opportunities and insight into the development of our health and lifestyles in general. The Province is working now with the Health Deal, but in essence, 9 out of 10 executive decisions still result in health reduction and pollution. This has already been referred to as the perverse reality of an era in transition, but it is certainly one to be taken into account. Resonating with health is a learning process in which many variables interact. Our AiREAS participation in the Marathon was our first chance to take a look at a totally new world and was a potential eye-opener for policy makers, citizens and entrepreneurs.

Enthusiasm grew within the AiREAS team, and our proactive "can do" attitude confused the commercially-oriented marathon organization more and more. AiREAS relates to the higher purpose of health, which leads us, and treats all partners, including the Marathon organization, as equals. As our ideas became more concrete, so did the demands on the organization for facilitation. Whatever we were given as support was gratefully integrated into our own evolution. The communication challenge was coordinated by Jean-Paul, and the medical one by Eric, while the value-driven entrepreneurial challenge appeared everywhere.

5.2 Communication Challenge

AiREAS has its own source of information that can be made relevant to the Marathon challenge: historical and real-time air quality information. Also, our accumulated expertise became relevant to both the Marathon organization and the Media. AiREAS reasons from a perspective of regional health and shared responsibilities. The Marathon and the Media had different interests of their own:

- Marathon organization: How does Eindhoven compare to Beijing? Can a focus on healthy air bring in top athletes who want to break world records?
- Media: A sensationalist approach based on highlighting possible negative (the city council discussed the AiREAS call for a carless Sunday at the governance level) and positive (will we break a world record this year?) consequences of the AiREAS alliance.

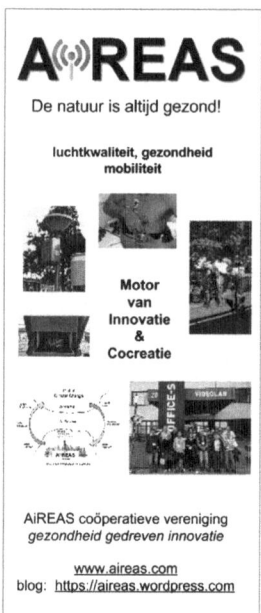

One of the information banners created for the Marathon

For communication, we again distinguish between information supply (making visible the invisible) and actual communication. The first is unidirectional and the second is interactive, with a focus on health involvement and measureable progression.

1. Informative

   (a) Air quality during Marathon 2015 in Eindhoven
   (b) Air quality comparison with Marathon Day in 2014
   (c) Air quality information from other areas of the world, e.g., Beijing
   (d) Medical health findings from those same areas
   (e) Innovative actions for healthy air (banners, blogs, etc.)
   (f) Informative interaction with the media.

2. Communicative

   (a) Communicative interaction with the media
   (b) Interaction with athletes and supporters
   (c) Interaction with visitors
   (d) Interaction with sponsors and organizations
   (e) Workshop (3rd of November)
   (f) The Brabant context.

Within these lines, we would be able to see if we could positively stimulate the people and positively affect the environment with measureable results. The communication team was developed for the purpose of interpretation and publication of specific Marathon-related air quality and health insights, and the accumulation of video material and interviews and interaction with the media. The team consisted of:

Jean-Paul Close—supervisor, media contact and researcher
Andre van der Wiel—camera team, together with his sons
Hein Kuiper—CityTV
Jason Clarcke—Fontys student and interviewer.

5.3 Medical Challenge

Eric de Groot, together with his POP and professional team, set up a temporary lab. A call was sent out to the participating runners to get them to participate in medical research equivalent to what we had done in the POP1 earlier that year with the 40 volunteers described in Chap. 2. Again, the objective was to look at health and lifestyle aspects of the participants and try to relate this to the quality of the air. By lining up all research activities, the results could be compared.

There were also differences.

- The intake interviews were done by members of Eric's own organization.
- A German product developer, Dr. Lutz Kraushaar, used the Marathon to test his brand new software and method of charting the vascular aging of participants by applying only 4 non-invasive pressure points.
- Pierre Cluitmans had acquired TomTom watches, so the runners would each carry one during the race.
- Pierre also enlisted a university student studying electrotechniques to assist him with the HRV measurements.

The medical research team ultimately consisted of:

- Dr. Eric de Groot—supervisor, assisted by 2 members of his team
- Dr. Pierre Cluitmans—HRV and TomTom watches, assisted by a student
- Dr. Lutz Kraushaar—vascular age charting, assisted by his wife.

Recruiting participants was easier this time because of the context-driven alliance with Golazo. A call was made through Golazo's e-mail list and the response was instantaneous and quite good. We even had to disappoint some people because our capacity of 20 individuals had been fulfilled. This proved to us again that connecting people through context is much more effective than cold calls that have no multiple interaction on a personal level. What does this mean? It proved much more effective to address combinations such as sports/air quality, environmental affiliation/air quality, study direction/air quality, physical exercise/health, sports/health/air quality, etc., than simply air quality and health. People tend to connect their own To Be and To Do selectively, through a concrete, meaningful interrelationship between the two. The reciprocal "what's in it for me?" becomes relative to the reward coming from the To Do, not necessarily the moral To Be. This would lead us yet again to the conclusion that the issue of "health and air quality" is a regional leadership issue, managing the societal To Do impulses in relation to the To Be leadership. On the other hand, this would also mean that polluting commuter cultures formed out of the socio-economic interests of labor can be attributed to a lack of core value-driven leadership focused on managing the wrong culture and priorities. Linking communication to sport, health and air quality was the right type of leadership, with the Marathon serving as the podium.

5.4   The Day of the Marathon

The announcement of the alliance between AiREAS and the Marathon had already drawn the attention of the Media. To our surprise, we found that the finish line had been built right at the location of one of our measurement stations. In terms of media attention, this was excellent, especially because it was also on the doorstep of the hotel where all the sponsors had gathered.

The Airbox had a prominent position

During TV and radio interviews, I experimented with messages that called for attention to air quality in relation to the runners. The call to avoid BBQ-ing the evening before and the suggestion to introduce a "carless Sunday" to honor the event had particular resonance. This became apparent through the many personal reflections I received from people in my own surroundings who recognized me from the TV appearance, as well as sports people who mentioned the call when interviewed during the event. Later, Golazo came to us with the curious feedback that one of the major sponsors, DLL, had cancelled their large sponsor barbecue and had organized something else to entertain their guests. All these lines of feedback showed the power of purpose-driven, event-related communication.

On the day itself, the response to this type of communication was even greater than it had been during the week. This seems logical from the perspective of context. When an interview with a "leave your car at home" message is heard while someone is commuting to work, the resonance is different than it might be while supporting your son, daughter or wife while they engage in extraordinary physical exercise. The moment in time and the circumstances at that instant also determine the degree of individual perception and receptiveness to information.

We also experimented with banners and stickers to visualize our presence during the Marathon, but time was too short to get our message well-integrated into the organization. The Marathon as an event is still set up from a business perspective and never from the ideology of a contribution to societal health. Co-creation was limited to the level of awareness of the challenge and the willingness to co-invest. Golazo offered their channels and infrastructures and AiREAS did their own thing.

Organizing such a big event is an enormous effort, involving 23,000 athletes, 200,000 spectators, 400 reps from sponsors and 1500 volunteers.

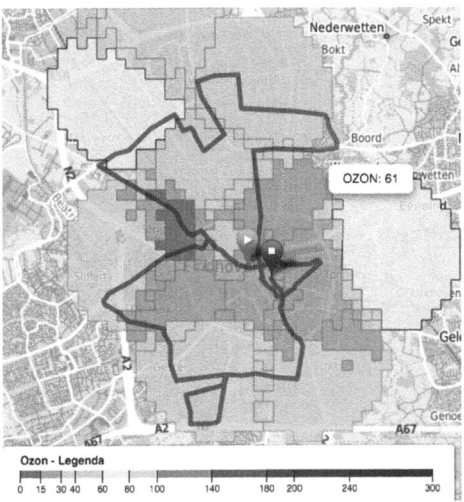

This image shows the Marathon route through town with Ozone levels mapped across the image

AiREAS's higher purpose could not take the lead at this stage, since the entire organization had already been set up. We had to resign ourselves to whatever we could get in terms of space and integration. Golazo gave us a spot in the hotel among the sponsors to tell our story. For the medical research, we received a number of square meters in the exhibition hall where the runners would gather before starting. With great creativity, the AiREAS medical team installed themselves and interacted with this environment, also a totally new experience for us.

5.5   Air Quality

The Marathon day itself was clear and sunny, with a blue sky, an excellent temperature for sports and a healthy start from an air quality point of view. It seemed that the call for attention to this point of view (no cars, no bbq's) had worked out well. This is impossible to confirm, of course. It is already difficult to move around town with a car when the major part of it is blocked for the Marathon. The verbal discouragement is probably a minor factor compared to the practical discouragement of the road blocks.

From PM10 and 2.5 perspectives, we see that the day started out with an average of about 25 micrograms per cubic meter ($\mu g/m^3$) and 20 $\mu g/m^3$, respectively. During the day, this average would lower to 17 and 10 (see figure below).

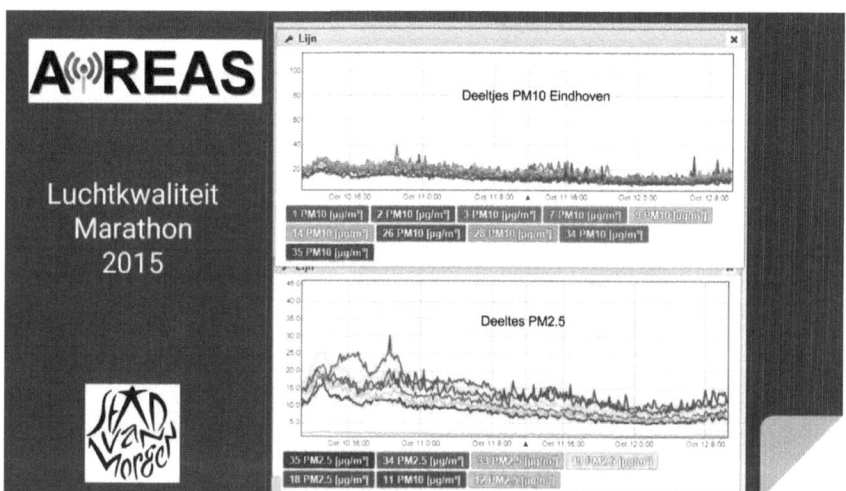

Air quality on October 11th (Marathon day). Graph on top is PM10 and on bottom PM2.5

Ozone peaks in weeks before the Marathon and on the day itself, in Eindhoven and Breda
(benchmark city without a Marathon that day)

On the left hand side of the picture, we see the relative Ozone levels in the weeks prior to the Marathon. Various peaks can be observed related to heat spills. The city of Breda has just 2 airboxes, which show similar but relatively higher peaks. The problem we face in analyzing the ozone information is that we have doubts about the absolute levels registered, due to the difficulty our technological partner confirmed in measuring these gases. So, we have to stick to the relative information of fluctuations.

On the Marathon day itself, we can see (right hand side, middle picture) that the ozone level was relatively low at the beginning of the day and built itself up gradually throughout the day. Intelligence shows that ozone is a very unstable gas that builds up from other substances in the measurement field of PM as temperature rises. We observe, then, a decline in PM and an increase in ozone. This was the case during the Marathon.

Comparing the data with that of one year before (2014, the top picture on the right hand side), we see sharper peaks and a longer low period. This was caused by a different type of weather in 2014 than that in 2015. Looking at data without context is, hence, crazy and delivers hardly any workable insight. It is difficult to imagine how ministries have dealt with policy-making by simply looking at data and averages when context is the most significant factor for getting an idea as to what is really going on. It also strengthened our idea that established norms of pollution hardly say anything about the wellness of a region. The context-related exposure of individuals and localized circumstances say much more. It is thus not the fragmented air quality norms that are relevant but the health of our citizens in their daily socio-economic dynamics.

It is hard to imagine the Marathon in Beijing and compare it with Eindhoven. Beijing has concentrations of PM that are, on average, up to and above 10 times higher than those in Eindhoven. Chinese authorities that visited AiREAS laughed at us for making such a fuss about our air quality. "You have no problem", they tended to say, looking out of the window. When we explained that our focus on health rather than pollution has become the newest innovative driver influencing our social cohesion and entrepreneurial spirit, their attitude transformed into one of curiosity. From a carless Sunday action in Brussels in September 2015, we learned that such an initiative has an immediate temporary effect on the air quality in town, but Beijing is probably not just polluted by traffic. All the measures taken by the Chinese government to host their own Marathon or Olympic Games may produce a short period of improved health, but we had a different, non-remedial approach. In Eindhoven and North Brabant, we wanted to see how sports, physical exercise and air quality could enhance our quality of life and productivity all year round.

5.6 Ozone

The geographical conditions in Eindhoven are good for a marathon race, and if air quality can be positively taken for granted, then the race could become one of the fastest in the world. This year (2015), it registered as number five on the world ranking of speed. Various people deliberately choose this particular marathon in their quest for personal achievement. And a number confirmed that they did achieve personal records. Curiously, as the ozone increased during the day, more complaints were heard by athletes who did not achieve their expected personal record despite the beautiful day, while in the morning, when the ozone was low, the opposite trend was observed. Ozone is a gas that irritates the lungs, so the observation makes sense. The feedback was, however, subjective, and involved too few participants to make it scientifically valid. Logically, we may be able to get better feedback if we start concentrating on this particular issue based on this well-founded suspicion. This became something to think about in regard to subsequent events and next year's marathon.

5.6.1 Conclusions from a Communication Perspective

The link between the marathon and air quality triggered a lot of interest among the media. This helped develop awareness, especially among the people who resonate with sports, and this event specifically, for personal reasons. Interestingly, AiREAS was approached a number of months later by people who said that they had left the car at home and avoided barbecuing to support the athletes. But they also expressed their frustration at having gone to such trouble only to observe a new event in town a week later, Dutch Design Week, in which old-time buses were used to transport people between regions in town. Those buses were highly polluting and demotivated the people who had shown goodwill the previous week.

This shows that we still have a long way to go. The perverse situation of a town with an executive health deal that still makes 9 out of 10 unhealthy decisions continues. But we need to start somewhere, and we need to show continuity, determination and celebrate our progression. The marathon link has provided us with multiple media appearances, exposure to huge amounts of people and excellent new insights that we can develop into roadmaps for health. Our own camera teams on the street also produced lasting material that demonstrates the psychosocial evolution people go through.

Among the marathon sponsors, we also found business people who were orienting themselves towards becoming much more involved in the value-driven progression and 4× profit philosophy of STIR and AiREAS through the Pyramid Paradigm described in Chap. 4. This also shows a trend towards business-oriented entrepreneurship in which profit becomes relevant to value creation, rather than speculation. Without much effort, we united over 8 of such value-driven innovators for follow-up projects and new initiatives.

5.6.2 The Medical Research

Our temporary lab was set up in less than one hour by Eric's team. The experience of the POP in the first half of 2015 showed. People interacted with each other instantly and trustfully. The first athletes showed up and could be dealt with immediately in a well-coordinated chain of research events.

The temporary research lab

- Intake interview
- Vascular charting
- Cardiovascular measurement
- HRV
- Handing out the watches.

The athletes in this research were of different ages. They did not all run the entire marathon, but did all participate in an intense race for which they had trained a lot. Comparing the different communities and their health characteristics, we could obtain some interesting cardiovascular insight, even while dealing with relatively small groups of people. The information had been gathered from the POP and Marathon, in which we also distinguished between people who smoke (air pollution) and those who don't.

The results measured in vascular wall thickness can be seen in this graph, drawn up by Dr. de Groot and his team. The information is clear. People who do regular physical exercise, such as training for the marathon, showed a slower vascular aging process than those who don't. People who smoke show a much faster process of thickening, which makes them much more vulnerable for strokes and attacks.

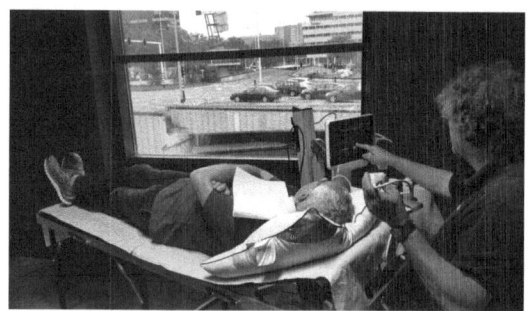

Dr. Eric de Groot doing his cardiovascular research

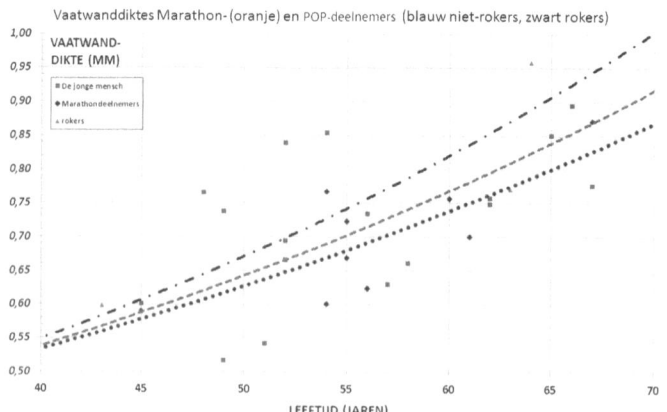

Aging of the vascular wall (thickness in mm) by age
Orange = Marathon participants, Blue = POP participants, Black = smokers

The information gathered by Dr. Kraushaar shows that sports people have a higher stroke volume than other people, meaning that their heart pumps more blood per stroke. Thanks to their sporting activities, their hearts display chambers of larger volume. In general, they would need fewer strokes per minute to oxygenate their system.

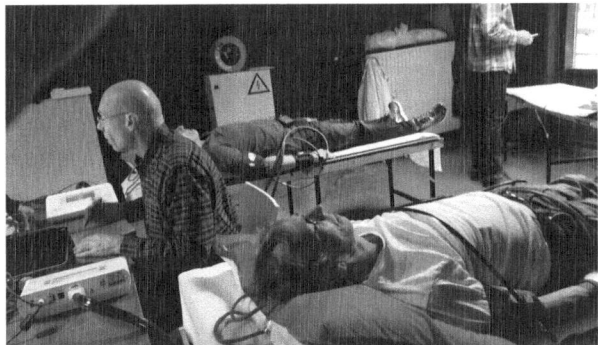

Dr. Lutz Kraushaar with his unique vascular mapping techniques

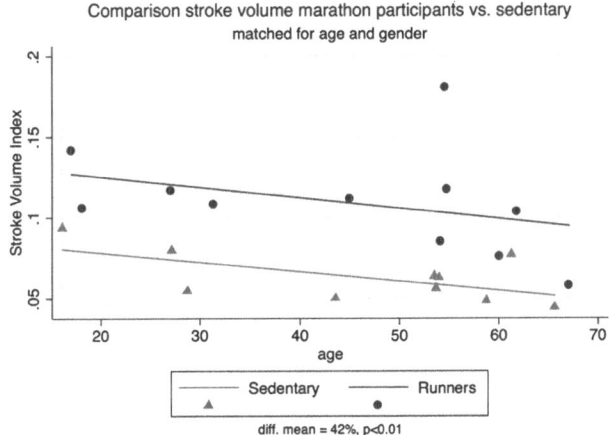

Also, vascular flexibility is much better among runners in comparison with non-running people. We had called for two persons from the same family, one a runner and the other a close relative. We could thus assume that their lifestyles and living conditions were as similar as can be, allowing us to relate the measured differences primarily to sports. Of course, all kinds of other factors could have been influential, such as DNA, job stress, professional activities, etc. But seeing the results, we can very confidently state that sports and physical exercise have a determining effect on the health of the heart and arteries.

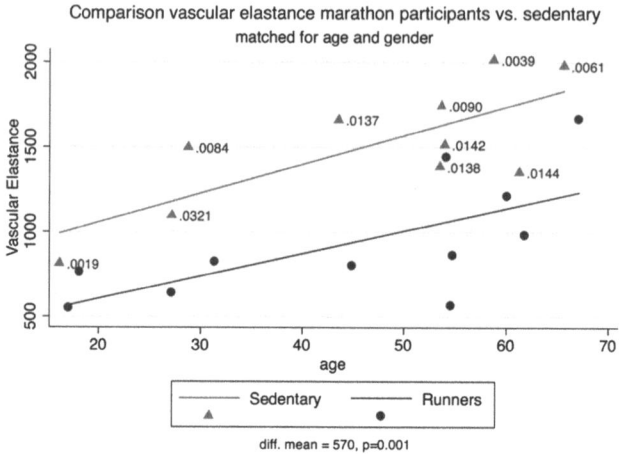

Pierre Cluitmans and his methods examined Heart Rate Variability both when at rest and while engaging in the running exercise.

Dr. Pierre Cluitmans doing his HRV registration with people at rest

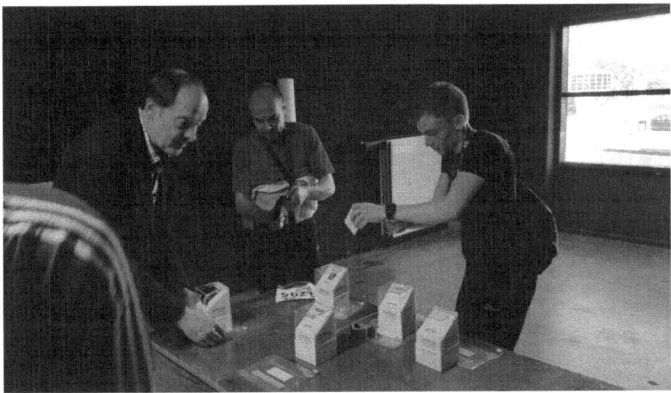

Dr. Cluitmans collecting the HRV registration devices from the athletes who came back

As we have already seen in Chap. 2, the amount of data collected on Heart Rate Variability is tremendous. The impressive set of data per individual, placed in the context of lifestyle or sports, gives very valuable information about the way a person deals with stress and stress recovery leading into new phases of rest. It can also be seen to what extent sports influence such variables.

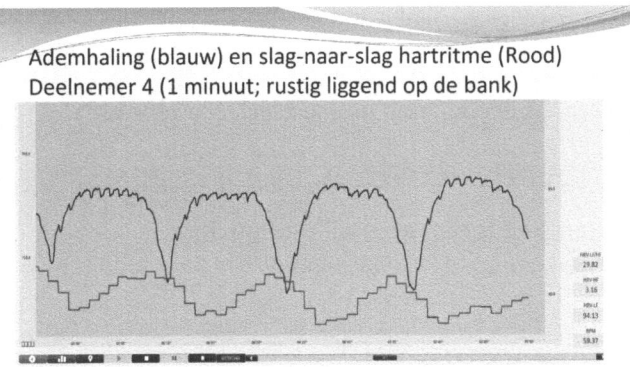

Heart Rate Variability (Red) and breathing (Blue) of a participant at rest
1 minute laying down on the bench

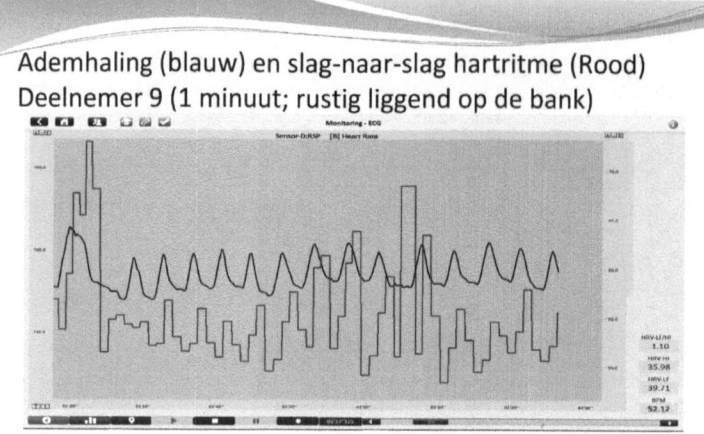

Breathing (blue) and heart rhythm (red) of another participant at rest
1 minute laying down on the bench

We can easily see through heart rate and respiration that every person is different, which makes such an investigation so interesting and special, but also so complex. We also see that there is an inverse relationship between respiration and heartbeat. When engaging in sports, this is, of course, important too. Interestingly, if you compare this image with the one on variable economics, the resemblance is striking.

The data that Pierre collected on the runners during the race shows the relationship between the development of speed and their heartbeats.

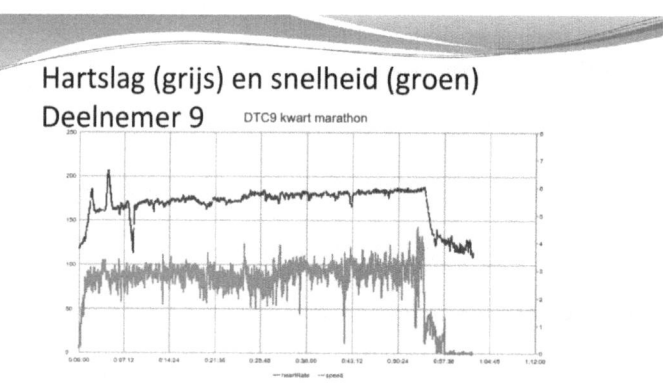

Example of measurements of heartbeat (gray) and speed (green) of a participant during the exercise

5.7 Conclusion of the Medical Data

Detailed analysis of all this data can zoom in on the health situation of a particular person and even predict possible heart or vascular problems at a later stage in life. The technology and research options are powerful instruments for preventive care and a stimulus for the evolution of a healthier lifestyle through making visible the invisible, including at the level of our own life support systems. In this way, preventive care can save the community enormous healthcare costs while enhancing the overall regional quality of life and the way people interact with each other and their environment.

In the book (AiREAS Phase 2) written by the health scientists and researchers themselves, more detail can be found about their proceedings and findings during the POP process. The link to that work will also be made available through Springer and our communication channels.

5.7.1 Overall Conclusion of the Marathon Exercise

Large events engage people with a particular mindset, allowing them to resonate more with related subjects than they might otherwise in daily life. The marathon showed the link between sports and our health. As an event, it had great communicative value for creating a health-driven mindset and pinpointing problems that otherwise remain largely unspoken of within the current paradigm of political-economic steering. From a research point of view, it was extremely valuable to be able to compare the data of different groups of people in society and show the relevance of physical exercise at all ages. The comparison with people who smoke also shows how lifestyle choices affect the health perspective negatively. While smoking is a personal choice, exposure to outdoor air pollution is not. In some reports, the authors compare exposure to outdoor pollution as being equivalent to smoking between 4 (Eindhoven) and 40 (Beijing) cigarettes per day. Applying this analogy to the POP research, it shows that taking responsibility together (citizens and governance) for the quality of our shared air is directly related to the evolution of our health.

Chapter 6
Multicultural Context

Jean-Paul Close and Rüstem Demir

Not many cities in the world can be seen in the present day as being a community of a singular cultural background. Most have seen migration patterns from within and outside their national boundaries. Eindhoven itself grew instantly to 40,000 inhabitants in 1920, when the small transit town of about 6000 souls annexed 5 villages so as to gain in size and extension. This was necessary to host the many people who were attracted by the industrial activities of Philips (lighting and electronics) and DAF (automotive). The city grew rapidly, now counting over 220,000 residents. The workforce needed exceeded the capacity of the Dutch community, so borders opened up for guest labor from other countries. Post-war development saw lots of people from Southern Europe, and later migration came from Northern Africa, Turkey and the old Dutch colonies. Right now, Eindhoven counts over 150 different nationalities. The largest sub-communities are Turkish and Moroccan.

When we decided that we wanted to work on a healthy city with the innovative participation of the local population, we also realized we needed to find ways to address the cultural diversity. Since the credit crisis, a lot of people had experienced some sort of isolation due to their foreign origins, looks and lack of integration. In times of economic abundance, such individuals could well take part in economic activities and still remain socially isolated from the main culture. When I was asked to speak during a set of encounters designed to highlight the complexity and value of a great diversity of cultures interacting in one region, I could draw from my own extensive personal experiences as a global expat executive between 1981 and 2001, research performed in Nordic countries in the '90s by my old Swedish teacher,

J.-P. Close (✉)
STIR Foundation/AiREAS, Sustainocracy, Eindhoven, The Netherlands
e-mail: jp@stadvanmorgen.com

R. Demir
Stichting Bij de Tijd, Eindhoven, The Netherlands
e-mail: rustemdemir@hotmail.nl

© The Author(s) 2016
J.-P. Close (ed.), *AiREAS: Sustainocracy for a Healthy City*,
SpringerBriefs on Case Studies of Sustainable Development,
DOI 10.1007/978-3-319-45620-1_6

Anita Ekwall, and my multicultural marriages. The Nordic nations went through an economic crisis back in the late '80s while hosting Chilean refugees fleeing from the dictator Pinochet. At the same time, Sweden and Finland were exchanging industrial activities through takeovers and fusions. Both on the level of company boards and in the streets, there was intercultural misunderstanding and unrest. Research by people like Anita Ekwall on the basis of local boardrooms and Geert Hofstede worldwide was published in subsequent years. These publications[1,2] helped to visualize cultural differences through pictures and my own research.

It became a new personal challenge to see how those other cultures could become involved in the healthy city challenge of AiREAS. What differences could be detected in lifestyle and how could the context of health and air quality trigger a response among this segment of the population? Contacts were established with multiple NGO's and individuals of different ethnic origins. Not many were interested in developing new kinds of social relationships, because they were being subsidized to perform other types of activities. Suggestions for addressing the different cultural groups through the health challenge were not seconded by the subsidy providers, because such activities did not clearly focus on creating jobs, economic growth deliverables or integration. Early participants from Africa, the UK, Morocco, etc., helped to set up some seminars dealing with 'The fear of the other' and 'Cultural diversity and sustainable entrepreneurship', but these did not connect proactively to the other communities; they just explained the difficulty of integration and the beauty of cultural differences and interaction throughout history as seen through language, architecture and cultural development. Our communication through the media and blogs was in Dutch, while many the subcultures tended not to access such media due to language difficulties and lack of contact. It had become a major, mind-boggling dilemma figuring out how to connect all cultures to the higher purpose of health and air quality, not just the mainstream Dutch.

6.1   Erasmus+

For several years, the high school teacher Rüstem Demir, who came from a background of Turkish migration, had joined the evening inspiration sessions of STIR academy. He had learned about Sustainocratic views, but found it difficult to apply them in his current job, as a teacher in a Dutch school for professional education. Then, he was contacted by a Turkish agency that was looking for partners in the Netherlands to develop the European Erasmus+ exchange program for students. He contacted STIR to see if we could incorporate some of our views into the inspiration sessions for those visiting youngsters from Turkey. We rapidly linked the two challenges and tried to get the Turkish community, the largest

---

[1]Mötet (1999) Anita Ekwall, **ISBN** 9789525306125 Ad initium **Utgivningsår**.
[2]Hofstede (1991) 'Allemaal andersdenkenden'.

foreign presence in Eindhoven, to open itself up for visits from students from Turkey so they could learn about AiREAS and the healthy city challenge. To our surprise, this time they accepted!

When asking this group of foreign, long-term residents in Eindhoven "why is the segregation between cultures persistent?", the answer was directly related to the social, political and economic environment into which these people had arrived in the Netherlands. The following factors were given that stood in the way of the integration of cultures:

- Low level of education of the first generation of immigrants,
- Their fear for discrimination due to 'being different',
- Language problems,
- The persistent idea that 'we are here temporarily and will go back one day'.

During times of economic recession, this feeling was exacerbated due to local attitudes against foreigners, who were perceived to occupy jobs or take part in the social system, thereby reducing the chances for the locals. The Dutch have certain words for locally-born people and others for those who immigrated or were born from immigrants. Such linguistic classification is itself a form of discrimination, one that can also be found in the behavior of the system, government regulation and an overall culture around "the fear of the other" (an 'us and them' culture). This cold, materialistic, classification-based manner of dealing with people in an unequal way had been one of the main reasons that I had founded STIR, out of a powerful desire to reform and redefine a new society based on equality, respect and trust. That was also the essence of Sustainocracy, and for the first time, it could be applied to different cultures. A multi-faceted experiment started which introduced innovations into the field of intercultural involvement, education and persuasive communication, with both local and international components.

6.1.1 Turkish Students from Turkey

When the first group of students arrived, they were asked to help explain the AiREAS healthy city challenge to the Turkish community in Eindhoven. With this exercise, a variety of objectives were covered. The students were inspired to come to understand the complexity of air quality in relation to human health. They also learned about intercultural complexity and were challenged to use their creativity and study material to communicate in public. And they did all this by establishing teams that would each work out their ideas together. The students were all aged 15–17, and reacted to the opportunity to participate with great passion and drive. During their preparations, they were supported by professional teachers who helped them forward with the tools and understanding they required along the way. According to the visiting teachers, the groups of young people learned more in four days through their self-educative impulses than they did in one year at school at home. The fact that they were in another country, challenged with a real life issue connected directly to their own awareness and talents, had opened up their minds and creativity, and their ability to perform with joy and happiness. Interestingly, after the exercise, we received calls from parents in Turkey who wanted to know what secret we had used to motivate their children to such a great degree that they had started to learn a foreign language and paid much greater attention to their studies!

A new educational theory developed that was equivalent to the practical proof of principle of the Transformation Economy. This type of economy positioned core value-driven change at the core of a productive, self-reflective community. This focus on change had so many innovative deliverables that each could show patterns of growth, producing continuous new economic development and impulses. The trick was to avoid focusing on growth, but rather learning to take growth for granted as a natural process if the value-driven change produced valuable innovations. The same appeared in the educational process. In asking these students to connect their inner motivation and creativity to a core human issue of perceived and recognized importance, their inner stimulus to search for knowledge and develop abilities was triggered. In giving them freedom and access to facilitating expertise, their learning curve jumped to unprecedented heights instantly. In the end, their deliverables were excellent, serving multiple purposes.

All these direct and collateral benefits were very much appreciated and added up to our own STIR participative learning curve. In subsequent groups, we would be able to anticipate such benefits and try to enhance them. Meanwhile, we confronted one of the Turkish communities in Eindhoven with the motivated presence of the youngest generation from Turkey carrying an important message. The students had put on a small show and entertained the Turkish people present with their youthful, modern enthusiasm and powerful message. The connection was made, bridging the local Turkish community with the local health challenge. It became clear that the core values of Sustainocracy are a common denominator for all cultures and serve the beautiful inter-human function of co-creation and connectivity, independent of language, religion or background.

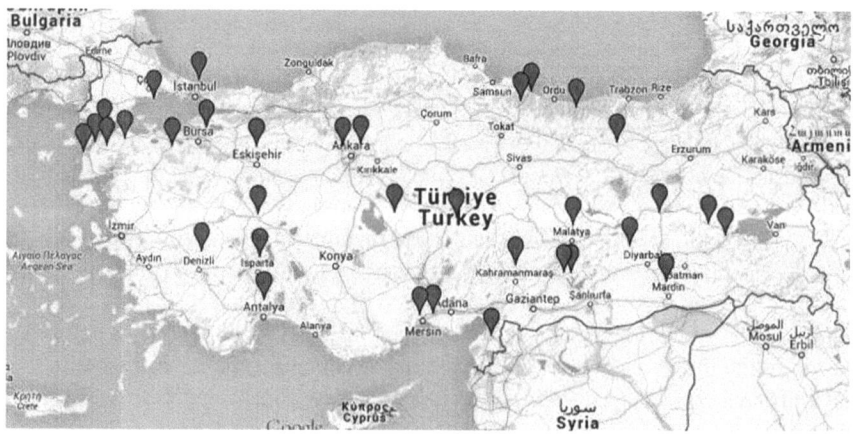

A view of the origin of the groups of students received throughout the program in 2015

6.1.2   Bridge to Turkey

Another curious side effect of the integration of the Turkish Erasmus+ program with the AiREAS objectives was the bridge function with air quality in Turkey. The students and teachers were also naturally looking at their own situation at home. Internet research rapidly revealed the local hot spots in their home country, making the challenge even more real. Posters were made that could be used to communicate with their local authorities and teachers started to integrate core values into their own programs and practicals for the students. Soon, politicians and government people started to join the exchange program and came over to get inspired by Sustainocracy. Not only did AiREAS become an issue, but so did the other STIR programs, such as the STIR academic method of participative learning and FRE^{2}SH local consumption and productivity. Local regional topics were discussed from the point of view of core values, including the problems the local executives face as the result of historical decisions that are now an impediment to progress and change.

One group, for instance, came from the small town of Soma. Soma is a town that was built around a coal mine. This mine fulfils a key role in the existence of the town, its local labor and the financial wellness of the resident population. In 2014, disaster struck when over 300 were killed in a mining accident. The group of visiting students that researched their AiREAS challenge on the internet was nearly all girls around the age of 17–18, young women on the verge of adult fertility and ready to develop formal relationships. They were shocked to discover that children currently born in their own home town have a life expectancy of just 50 years, due to the pollution created by the mine. Government officials said that they wanted to address the problem, but were blocked by a coal supply contract with a foreign nation. This nation pays for the coal in a competitive environment, but does not take responsibility for the local social and ecological damage.

We see this type of tension between old commitments and desired progress everywhere in the world. Extensive dialogues were developed about the complexity of change and the path of building up something new within the scope of the new reality of $4 \times$ profit, while downsizing the old commitments and consequences.

Another example is the delegation that came from Malatya, a town in the Eastern region of Turkey. The group was chaired by a local governor. With our partner, Rüstem Demir, we had agreed to focus our 3-h session on core value-driven entrepreneurship. We went through the cycle of human complexities and the $4 \times$ profit approach. Together with Eugen Oetringer, who introduced a method for highlighting opportunities and ideas from the group and pinpointing difficulties in reaching objectives, we were able to define the preconception of a health deal for Malatya.

Co-creation in action

One of the issues raised was the proximity of the warzone of Syria. A stretch of land over 250 km long was filled with landmines, a serious hazard for the local community and a conundrum for the local executives. Within the entrepreneurial network of STIR, we have a creative group of entrepreneurs who have developed various innovative initiatives to detect, map and destroy landmines in a cheap and effective way using drone technology. Their inventions have captured the interest of the United Nations. Their lab was just 5 minutes' walking distance away from the STIR academy, so a link was rapidly made.

Such issues were raised in many of the local groups, resulting in formally-stated intentions to use sustainocratic level 4 regional development to address these issues for the sake of sustainable human progress in each region. The visiting generations, both students and professionals, reached a level of understanding of the benefits that core values have for their community, the potential for tourism and foreign investment opportunities, and the evolution of wellness, social cohesion and pro-ductivity of the local community. The Erasmus+ program, as of this writing, still has 6 more years to go and promises to be a powerful instrument and platform for leveraging both Dutch and Turkish value-driven activities and relationships at humanitarian, educational, ecological and entrepreneurial levels. During the POP, we received over 700 students and about 70 teachers and regional executives. Throughout the program, we expect to inspire a total of 7000 visiting students and professionals, all while establishing and maintaining productive relationships with the local schools and communities.

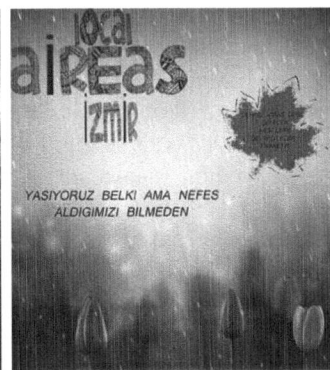

Some of the creative results of the students from Turkey

The program is, of course, not exclusive to Turkey and the local Turkish community. It just so happened in the course of the POP that we were able to experiment intensely with that particular social reality. The learning curve is now available for the entire world in the form of value-driven participative learning at all age levels using core values as bonding instruments.

6.2 Conclusion

We learned various things throughout this program. The first conclusion is that core human values are common for all people and have an effect of bonding and community building irrespective of cultural background. People overcome their differences with ease when invited to work together on the development of those core values. The different cultural backgrounds subsequently become an inspirational diversity of creativity. The Dutch tend to be more rational in their expression, while the Turkish lend emotion and colorful creativity. Language issues are overcome through images and illustrative videos on the internet that everyone can understand.

We learned to distinguish between information supply and the art of communication. How one gets a message across is different from how one gets people involved.

We also developed the STIR participative learning pattern. For over 100 years, we have educated our young generations by the standards of industrialized processes and labor in hierarchical structures. Obedience and mass productivity (no individual identity or initiative) were the leading educational drivers, as they still are today. But the enormous challenges that the world faces no longer require robotized human beings, but rather participative, passion-driven, creative contributors to progress. This demands a new educational system which STIR defined within the scope of our level 4 sustainocratic participation society. At various times, we have tried to introduce this way of thinking into the Dutch school system, but without real success. Every time we managed to introduce a program, such as our 'In search of the hero' or 'Entrepreneur of your own life' programs, the schools would embrace them at the educational level but not at the executive. When an executive commitment for continuity was needed, the doors would close and remained closed. The dependence on money, with national governmental guidance through inspections, blocked this progress. It was therefore very interesting to see the positive results within the Erasmus+ program with the Turkish students who were the first to really benefit from this type of learning process.

The impression arose that structural innovation never comes from within a system but needs to be developed outside first. We had already called this the STIR loop (see figure below), which was applied in AiREAS at the operational level. It takes executives outside of their regulated comfort zone into the core value-driven reality of Sustainocracy. The sense of freedom opens up the creativity of the executives, who tend to commit to the innovation they themselves propose. Subsequently, they address the burden of their own organization and help it to transform step by step through powerful arguments. However, the educational system is not governed by the executive of the schools, but rather by the financial dependence on the central government. The executive is simply managing the national interests, which are confined to the historical context around which national financing is based. Many young people feel a separation between educational obligations and what they themselves and society expect from them.

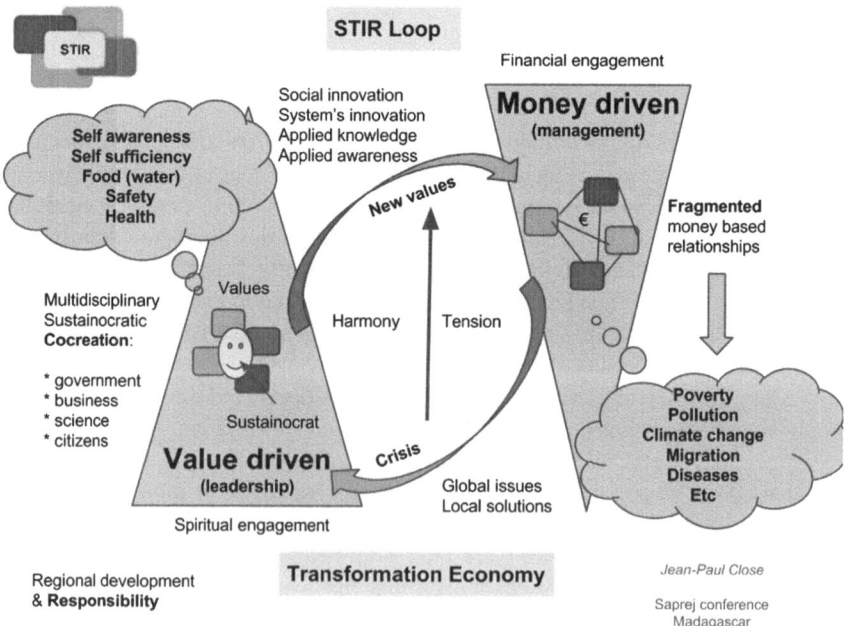

After our success with the Turkish youngsters, we also invited the Dutch schools to participate, and we gradually saw steps being taken in this endeavor too. Finally, the approach we had taken had become a true ambassador for ideological and practical connection with Turkish regional development. We started to consider opening up our program to other European countries as well. The key was in developing relationships of trust with local partners, and that would take time.

Chapter 7
Level 4 Sustainocratic Regional Development

Jean-Paul Close

In autumn 2015, we were surprised by two different but highly significant signs of support for our method of working:

1. The French multinational VINCI Energies gave AiREAS the European VINCI partnership award for innovation,
2. The research company VentureSprings positioned AiREAS at the level of a peer 4[1] regional development, the highest evolutionary level, in their report on Smart City initiatives in Eindhoven.

The combination of these two recognitions was a decisive factor for our plan to expand our sustainocratic format throughout the world. Sustainocracy was no longer considered a form of abstract idealism but rather a practical reality, not a revolution in the name of a 'different society' but an *evolution* towards a new, modern reality. Evolution is less confrontational than the disruptive change of a revolution and shows a pattern that we also can relay back to the Kondratiev sinus of economic evolution following the industrial revolution. The Kondratiev cycle had already been related to my cyclic pattern of human complexities, including a powerful reference to the musical string theory of Pythagoras and Galileo Galilei, with a clear link to the behavior of the entire universe. Human behavior in a free biological environment bonds in the same way that DNA molecules connecting to produce living structures do. Often, as a result, human beings remain trapped in old bonding systems and cannot connect to other energies to which they feel attracted, unless they free themselves first. This type of bonding into new value-driven communities has a different energy than that displayed by economic hierarchies and activities. They are parallel types of universe that interact but don't mingle. That

---

[1]Peer 4 has been defined by the Presencing Institute as self-aware global co-creation.

J.-P. Close (✉)
STIR Foundation/AiREAS, Sustainocracy, Eindhoven, The Netherlands
e-mail: jp@stadvanmorgen.com

© The Author(s) 2016
J.-P. Close (ed.), *AiREAS: Sustainocracy for a Healthy City*,
SpringerBriefs on Case Studies of Sustainable Development,
DOI 10.1007/978-3-319-45620-1_7

explains why economic lifestyles could become dominant at the expense of life on Earth. The system lacks the natural emotional and spiritual bonding mechanisms of awareness and choice that unite people.

The first economic boosts in the world were generated by new technologies applied to mobility and infrastructures, allowing globalization of industrialized trade to take place. The subsequent boosts were related to the petrochemical influence on productivity through plastics, electricity and individualized automotive abilities. Finally, the IT infrastructure of the internet and integrated communication facilities, making information accessible to the entire world, formed a bridge to the current situation of a strong global psycho-social evolution, the collective positive disintegration[2] that communities now go through.

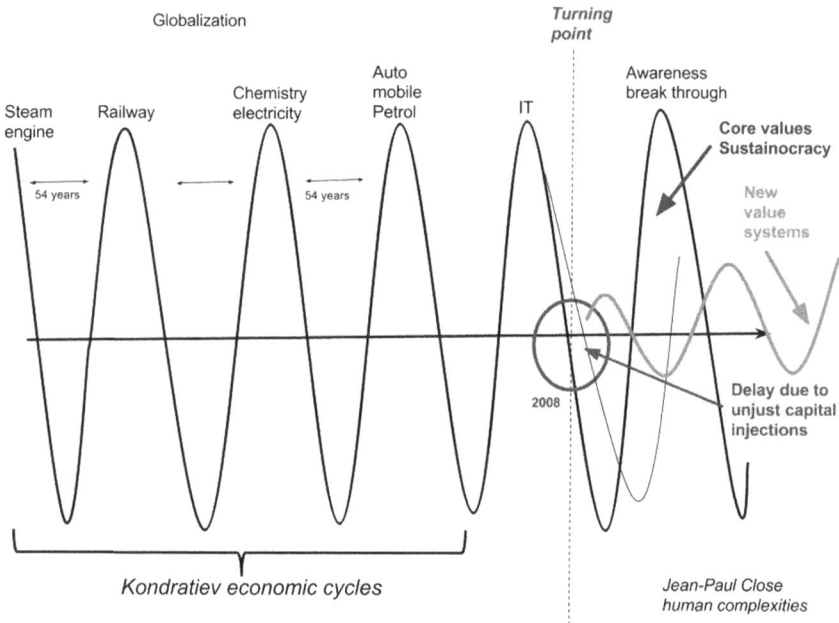

Huge changes are to be expected, since the credit crisis in 2008 opened up the minds of people to the obsolete nature of the current reality. This evolution is natural and contains a global psychosocial breakthrough. Global issues, local solutions, global application

In 2008, we could only speculate as to what would bring about the new Kondratiev peak, even though we ourselves had already experienced the deeper awakening of our individual awareness. The mainstream society was still managed by the old economic reality, which was kept alive through unjust capital injections. On the one hand, such injections were counterproductive, because they interrupted the natural cyclic pattern of renewal and kept the obsolete structures going for a

[2]Kazimierz Dabrowski: https://en.wikipedia.org/wiki/Positive_Disintegration.

while longer. The prospect of crisis would only get deeper and harsher as the consequences of the old global mismanagement of the political and economic hierarchies continued. On the other hand, it offered the opportunity for any alternative ideological and evolutionary patterns, such as Sustainocracy, to become more robust through experimentation, the grouping of new age initiatives and improvement of the argumentation through the build-up of proofs of concept. When disaster eventually strikes (again), then the alternative will have matured enough for sustainable progress. Evolution then becomes a genuine executive leadership choice.

The disaster of a crisis is now, however, not just economic, but also humanitarian and ecological. Numerous incidents around the world show that humankind is on the verge of collapse due to its own mismanagement and abuse. Entire cities have become vulnerable due to the effects of climate change and pollution, combined with the historical demographical positioning of the conglomerations around trade. The dependence on financial instruments and focus on hierarchies of materialism did the rest.

The major problem that cities and regions face is the hierarchical structure of our communities. The human motivation to be part of a community is generally limited to self-interest around economic facilities, social securities and services provided by cities. The general mental distance of any given population from our natural core values of life is unsustainably huge. We have grown used to relating to the dead things of materialism that surround us in abundance and to which we have given more value than our own potential creativity and the development of life. Our natural creativity, which made us stand out as a species with the ability to govern our needs through productivity and planning, has made us the 6th cause of a mass extinction of life on our planet since its birth, as we described at the beginning of this book.

Regional territorial management is based on the leadership of political and economic drivers that are consistently permitted to rise above universal ethics at the level of the citizenry. The economy of growth has become a norm, attached to the trade of death instead of the facilitation of life. Consequences such as health problems, behavioral disorders, pollution, and catastrophes are addressed as a costly problem to be remediated without any proactive actions that might avoid the humanitarian and ecological drama in the first place. The combination of generalized public apathy and reluctance with money- and power-driven political structures and fragmented, financially-measured responsibilities shows that we have reached an impasse that will be difficult to break through. This societal structure is not just obsolete; it is lethal, simply because no one can be pinpointed as being ultimately responsible for this disaster. In the end, it is our species as a whole that is responsible, but no one in our current context of societal complexity feels this pressure, passing the buck to others in the process. Democratic choices have mutated into an outsourcing mechanism of self-interest that has established a hierarchy around organized greed and power. In this field of perceived reality, change is not welcome.

7.1   Laudata Si

The drawing shown below, of the hierarchical structure between motivation, steering and ethics, was inspired by a theological and philosophical explanation after the publication by the Catholic Church of **Laudata Si**, by theologian and professor of philosophy and systematic theology, Eduardo Echevarria.[3] While the call for responsibility from Pope Franciscus I appeals on the motivational level through the meaning of 'Laudata Si' ('Praise be to you') for the care of our common home, this is a significant step away from the hierarchical positioning of the church for over 2000 years.

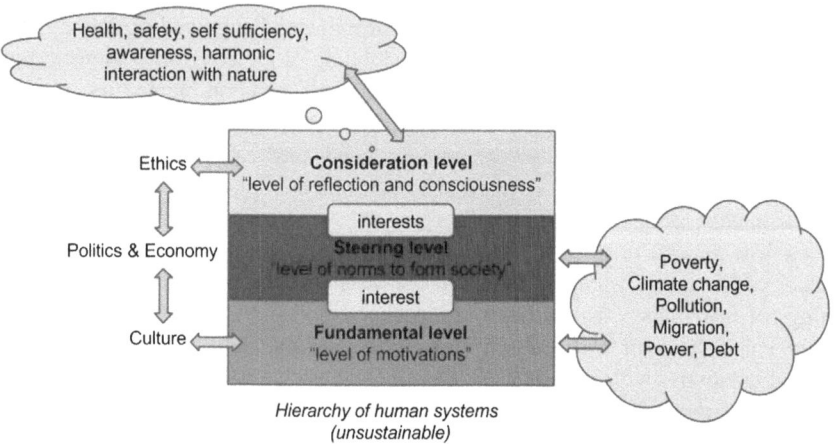

Hierarchy of human systems
(unsustainable)

The current failing system of executive regional development

In the tradition of hierarchies, the lower level represents the immense mass of people who have a low degree of awareness or even consciousness. They are being guided by those who are illuminated (priests), while the layer of ethics is in the confinement of God, interpreted by the directive of the human representatives who steer the people. This type of powerful organization has been copied by industry and is still in use within large multinationals and governments. Instead of guidance through the interpretation of God's message, it becomes political and economic. While there is room for reflection on the interpretation of morality (God's creation of life) in the church, there is none in the money-driven realities that worship the economized trade and possession of dead goods. Inspired by this ancient hierarchical structure, combined with my own awareness of psychosocial processes fostered in my work with STIR, and with knowledge of the same transition shown in the Kondratiev evolution, I modified the drawing in the following way:

---

[3]http://www.hprweb.com/2015/06/the-theological-mind-of-laudato-si/.

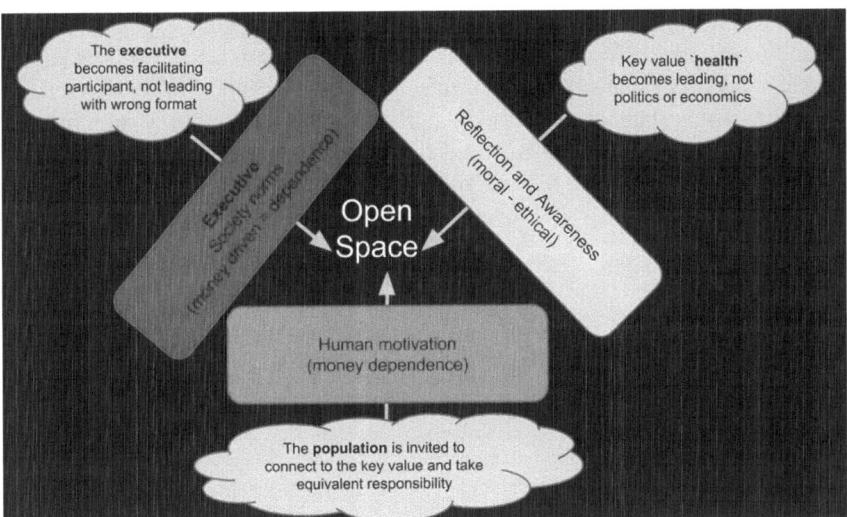

The executive
becomes facilitating
participant, not leading
with wrong format

Key value `health`
becomes leading, not
politics or economics

Reflection and Awareness
(moral - ethical)

Executive
Society norms
(money driven's dependence)

Open
Space

Human motivation
(money dependence)

The **population** is invited to
connect to the key value and take
equivalent responsibility

The fundamental change is that the core values become the leader and human motivation transforms from one of apathy to one of participation

The industrial hierarchies considered the development of knowledge at the motivational level to be important for enhancing their productivity. Education focused on cognitive issues, such as basic knowledge of language and calculus, within the normative of obedience to directives. Infrastructures were needed for the logistics of producing and moving goods, and this became the driver behind the Kondratiev cycles of economic peaks. The workforce was contracted to remain obedient to the industrial hierarchy in exchange for a salary and education. This educational requirement was taken over by the first constitutions to install the 'learning obligation' through the development of a school system based around the same principles of obedience and material cognitive knowledge.

People started to learn to read and write, gaining access to the wisdom of both the churches and the industrial economic processes. The reading of books by the population developed into the power for self-reflection, which brought people into direct contact with their own perception and confusion of moral responsibility. The psychosocial (r)evolution had started, awakening people to their own confusion concerning what life is all about and bringing them to the quest to understand right and wrong (Dabrowski's positive disintegration) at a personal level without the indoctrination of a church around guilt or an economy around debt.

Meanwhile, the globalization of industrialized processes had trapped humankind in a flow of blind self-interest, expressed in the desire to possess objects that give us a feeling of identity and differentiation. All attention had been taken away from life itself, away from the adaptability and creativity based on our core values of living life. The human world had entered into a devastating competition for resources in the name of sustaining the economies of trade and the individual desire to "have".

The industrial and economic hierarchies became dominant at the expense of global life and moral education. The church and its indoctrination tended to disappear as humankind found satisfaction in self-interest and a perception of abundance, not realizing that this abundance is based on producing death, not life. Within my own lifetime, the world population rocketed up to the astonishing amount of 7 billion people, further growing to 9 billion. This combination of factors has brought us to a point of singularity which will invariably lead to a generalized collapse.

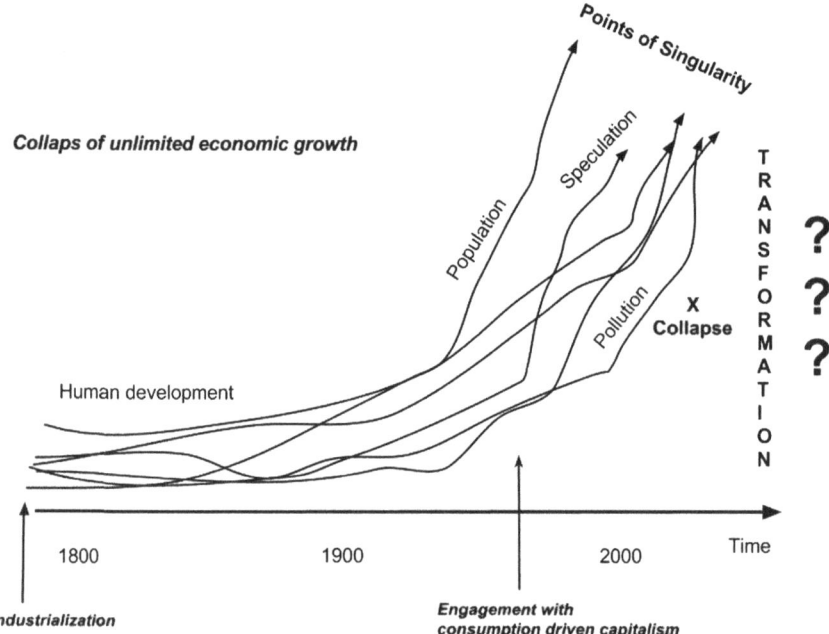

But this collapse has only to do with our materialism, not our original free human spirit. As this happens, our wit and creativity in regard to life is once again stimulated, developing a movement for change that has already crossed over the chaos the materialistic focus, allowing us to let go of that while concentrating on life. The human spirit follows natural patterns between relaxation and stress, as we have seen in our medical research into heart rate variability. But our economy only follows the single pattern of relaxation and has no mechanism for stress other than its own collapse. To compensate for this, a dual economic system needs to be put in place, one of both transaction (relaxation) and transformation (stress), representing continuous interaction within a community that can keep it alive and progressive. The next comparison shows how the human body works and how the economy could work in the same manner.

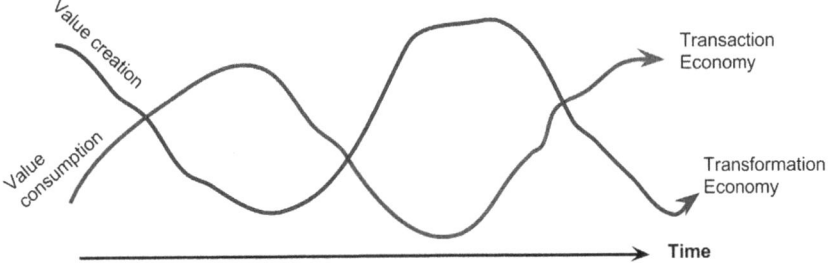

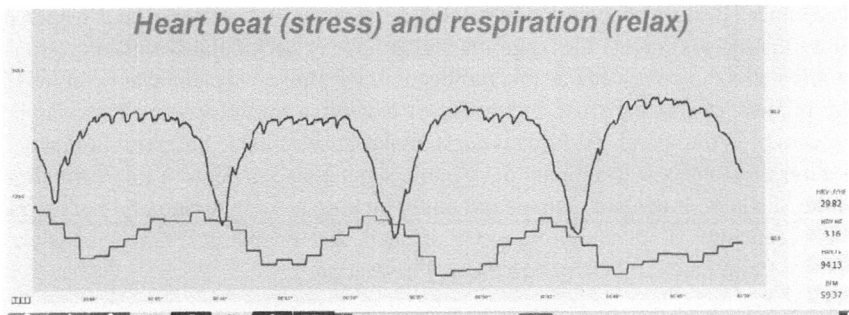

Comparing nature with economics

This is exactly the message and promise of the peer 4 level of regional development and the Sustainocratic multidisciplinary way of working. Globalization of materialism comes to an end and the localization of living mechanisms of co-creation are developed out of human self-preservation. This has huge implications for our societal formats. Peer 4 Sustainocracy helps this evolution forward, avoiding the collapse by creating a new workable format using the positive elements of the old system and adding new mechanisms of life.

7.2 The Big Turnaround

The regional executive is forced to step aside as large chunks of society start taking responsibility themselves, through the guidance of their own perception or interpretation of core human and universal values. Additionally, the credit crisis demands reactions beyond the political and economic hierarchy. The population starts demanding space to develop their new reality, as we have seen in Egypt and the so-called Arab Spring.[4] On many occasions, the old hierarchy tries to keep itself

---

[4]https://en.wikipedia.org/wiki/Arab_Spring.

intact, even at the expense of civil war or regional genocides. In nature, we see that a heartbeat elevates out of control, up to the level of a heart attack, if not compensated with rest through respiration and recovery. Death is then a serious possibility. Analogously, the economy will not survive if not oxygenized with fundamental innovation for change.

In certain cultures, the executive and the population become partners in seeking sustainable progress together. That is what occurred in Eindhoven with AiREAS. AiREAS created a new Kondratiev energy, a new social economic resonance. This compensates for the old, singular one.

Such events have everything to do with the people involved. In some cultures, the hierarchy of power is so demanding that a sustainocratic partnership is out of the question. In others, like in Eindhoven, both those at the executive level and the academically-educated population related more to shared authority and responsibility than to a systemic blockage maintained out of accumulated self-interest and manipulated directives. In our first publication, we already explained at great length the process of moving from individual to collective awareness and the complex process of letting go of old formats for the sake of sustainable progress. In practical reality, this means that regional development must also undergo a transformation. But essentially, it evolves. We are not really looking at a big turnaround but merely the acceptance of diversity into our societal and economic reality, a natural, self-aware interaction between stress and relaxation.

7.2.1   4 Levels of Regional Development

When we look at the development of cities over time, we can see many different types of motivator that have gotten a population to gather in close proximity to each other. The protection from hostile attacks by other groups was one such clear motivator, as was the existence of a marketplace for the trade of goods, such as at regional crossroads or in harbors. In all cases, the first level of regional development was the deployment of a basic infrastructure. This could be in the shape of a marketplace, a place where ships could dock, roads, etc. Housing facilities for those who remained around the facilities and actions were a logical consequence. As communities grew more complex, the infrastructures needed to integrate in order to optimize space and resources. The local human motivation of self-preservation was always the driver for this clustering, never core sustainocratic values. Why was this? Because our awareness had not yet reached a point of understanding. There was plenty of space to grow, and enough from which to take. This is not the case anymore.

As such, we learned to value dead products and imaginary financial resources more than life. The products became commodities with which we could show off status and wealth. Cities became conglomerates of the exchange of goods and public events. The impressive growth of cities over time, especially over the last few decades, shows how much time the local executives had to spend developing the first and second infrastructural layers of their regions. In the following videos, one can get an idea of the expansion of such places, which now house many millions of residents:

https://www.youtube.com/watch?v=2WGPvWPpey8&list=PLzYZm159uzQNc7H 5UCCXHx4c4TKdCeaNt.

The combination of an influx of people and the mentality of survival in an era of greed for those who had to deal with the city's financial reality made cities a hotspot of vulnerability, pollution and all sorts of illegal (and legal) criminality. The need for regulation and control was already visible in the early 19th century, but became even stronger when the consequences developed exponentially. With regulation and control, we see government grow in the responsibility and costs needed to address both infrastructure and behavior at the same time. Level 3 (Smart City) hence became an instrument for government officials to further extend their control systems and try to let technology help in planning the city's short- and long-term evolution. Cities had become very vulnerable due to their dependence on resources from outside the city itself that were only accessible through the power of the concentration of wholesale and the availability of money. Basic needs such as food, water, building resources, energy supplies, etc., all became economized, while jobs in the city mainly related to the distribution of goods, semi-government (education, police, health care, maintenance, infrastructure, control) and real estate. Increasingly, people were entertained with events and kept docile through social securities. At the same time, in many large cities, sub-communities developed within the chaos of poverty, greed and indifference, forming ghettos and social inequality.

Financially, a city needs huge amounts of money to develop, and this money comes from taxes, government services, sales of property, etc. But this is often not enough, and financial debt grows exponentially. Speculation develops into inflation, which becomes known as the economy of growth. This growth can then be taxed again, giving the relative impression that the city is getting richer, but at the same time, both the costs of society and moral (and real) poverty rise even faster. In reality, the region gets poorer and poorer, but the story is told differently by political and economic interests. It finally becomes important for the city that responsibilities be shared. In Eindhoven, after the recognition of vulnerability due to the dependence on large industries that had moved into low wage areas, the idea of the Triple Helix developed. The Triple Helix involves the three pillars of the old money-driven society: government, education and business development. These enter into cooperation in an attempt to optimize local resources, develop innovations together and implement Smart technological solutions to manage the city effectively. The human being is still out of the picture, as real estate, speculation and technology prevail.

7.3   Primary resource of a city

In all this evolution, there is a strong segregation between dead, hardware-oriented capitalism and the soft, humanitarian and ecological balance in the region. Cities had become enormous mountains of concrete, glass and cement, with huge levels of pollution that created human and environmental health hazards and destroyed productivity, as well as landscapes. Health care can flourish, but the anthropocene is well on its way. Poverty starts to rise and so does the discomfort of heart, lung and vascular problems, together with behavioral disorders such as ADHD, dyslexia, burnouts, autism, etc. The economic system uses the planet and humankind for economic growth at the expense of death. When looking at the presentations of Smart Cities, they will invariably deal with housing, infrastructure, technology, mobility, ICT, etc., but never with the primary resource of a city: the creative, self-aware, healthy human being.

7.3.1   The Choice

At the level of a Smart City, a choice appears. When applying smart technologies, what use do we give to the outcome and data? Do we use them to facilitate the executive system for the benefit of expensive bureaucratic regulation and control? Or do we involve the core value, the human creative power of the city, to deal with the challenges that we face? When we decide to do the latter, we enter into a new era, that of the participation society, a Sustainocracy, in which core human and natural values lead, triggering creativity and finding unprecedented innovative solutions for local resilience, self-preservation and wellness.

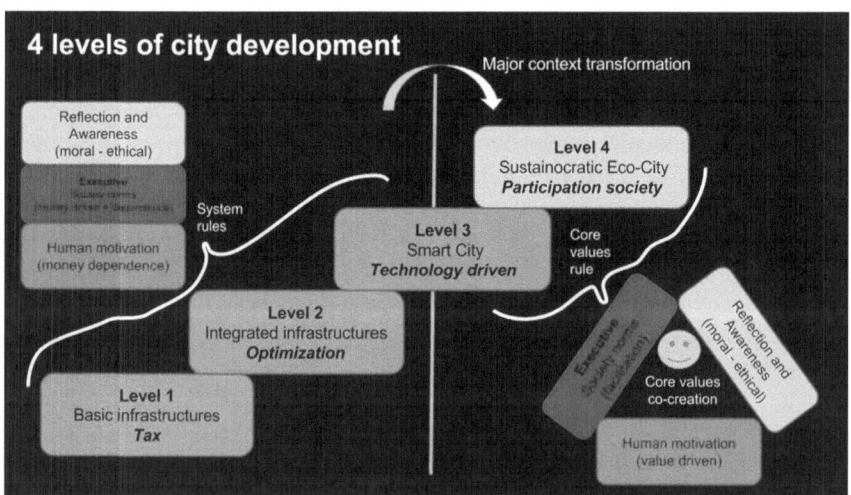

The context transformation is huge when deciding in favor of life rather than death

The rules of interaction at level 4 are totally different from what we experienced up through level 3. In the center of the open space in level 4 (the grey smiley in the drawing above), we bring together the 4 core stakeholders of a region, the fragmented institutionalized essentials of a human life: awareness, territory, creativity, knowledge and behavior. [AU: That's 5 things.] The core values of life become key, not the system nor the dead materialism. Our institutions become instrumental for serving life, instead of producing death. This places totally new demands on institutional and personal leadership, as we have seen throughout this book.

- The local government becomes a facilitating partner to its own population and eco-system,
- The local population becomes the primary resource for sustainable progress through value-driven creativity, co-creation and productivity,
- The local entrepreneurship becomes innovative within 4× profit dynamics as a partner towards a higher purpose other than product and service delivery,
- Education and science become servants in applying knowledge through participation processes, capturing and safeguarding new knowledge in the process.

The values created are shared, not accumulated individually.

7.4 The Sustainocrat

Key to this evolutionary process of reassembling the fragmented human characteristics of the old structure is the figure of the Sustainocrat. He or she is a representative of human life and its core values, independent of the systems and hierarchies. Without such a connecting individual, the multidisciplinary groups would become subject to the hierarchy of the territorial executive and ruled by the tradition of political and economic interests. When such an executive decides to work on the core values directly, they automatically come into conflict with the reigning materialistic interests. We have seen politicians get into severe trouble when trying to challenge the system from within. We need the leadership of such authorities, yet cannot sacrifice them to idealism. We have to find ways to use their personal awareness and empower their position by taking change outside of the system and then bringing it back in again. The afore-mentioned STIR loop holds the solution.

Executives that are invited to the value-driven co-creation tables cannot stay away as change occurs there. If they decide not to participate, they remain part of the problem rather than becoming part of the solution. This leads to a self-selecting process. Organizations that have no interest in change, nor the core human values, remain anchored in financial engagement while change occurs elsewhere. They have a tendency to disappear when change gets injected back into the system for the purpose of resolving the issues that upset the system in the first place.

Executives that step outside their financial engagement in order to commit to value-driven co-creation get the opportunity to look at their own organizations and the impediments to sustainable progress that the structure creates. Real leadership is then shown not only through proactive participation in producing change but also in removal of the obstacles that stand in the way of change. The position of such an executive in both environments becomes a transformative key. For the individual, this is a safe path, because obstacles are challenged using arguments based on proof, including expected results when complied with. There is no personal ideology anymore, but rather a core value-driven progress with practical instruments for change. Anyone in the old system who stands in the way of the progress of life can be dealt with through existing legal systems that claim human rights with respect for the integrity of life as their guiding principle.

For large enterprises, the challenge is to look at their participation in regional value creation as a feeding system or their own sustainable innovation platform. Rather than positioning their business on the side of economic growth through the massive exchange of materialistic instruments in a highly competitive environment, they can position themselves as catalysts for value-driven change with 4× profit expertise in those areas where 4× profit is also penetrating the local culture. This occurs everywhere around the world, often starting where the needs are most pressing and communities are small enough to enact level 4 co-creation.

Universal values
Ecological values
Human values
Economic values
Value driven cooperation
Trust
Respect
Equality
Individual talent/authority

Column of values
Sustainocracy - Close 2012

The most important effect of stepping up to level 4 multidisciplinary co-creation is the gradual appearance of the column of values that acts as a bonding mechanism between all participating parties. The old materialistic playing field had exactly the opposite column, producing segregation and disunity all of the time. This segregation and distrust requires extensive mechanisms of rules, contracts and control, while level 4 multidisciplinary unity can function without all those mechanisms. In AiREAS, no one is contracted nor hired; all interact on the basis of confirmed talent, expertise, equality, commitment and trust.

On a social level, this type of involvement was perfectly possible with people whose reciprocal wishes were related to other values (such as food, energy, electronics, housing, etc.) than those captured by economic steering (money alone). We have shown that those other values were related to participative education (persuasive communication, Erasmus+), awareness and lifestyle (backpack, POP1), authentic value-driven entrepreneurship (APP, Hackathon, ILM, ICT) or governance in transition (saving money, producing better cohesion, reducing structures, empowering authority, etc.). To access the masses, however, we need to introduce new value systems that can replace the old and be accepted by the providers of basic needs or at least involve those basic needs in the chain of value creation. People relate to their surroundings out of self-preservation. While the old economic system became corrupted and obsolete, it will remain alive, despite everything, until an alternative is accepted with which everyone can resonate. This alternative can be the direct involvement in producing one's own needs (such as FRE^{2}SH) or developing a new intermediary value system for sharing true values (such as the AiREAS coin).

In conclusion, the local deployment of level 4 regional development triggers the productivity of the population, but needs to be dealt with in a 'fair share' way through commitments, true values and reciprocity. A vibrant community is the result of passion, creativity and liveliness developing. People become aware of safe abundance through circular economies and social cohesion. Money is simply a means, just like natural resources, knowledge and creativity. The end result has to be harmony and wellness based on respecting the core values while eliminating greed and self-interest through the provision of co-created abundance for all.

The dialogue and societal rituals change as this new resonance with core values becomes accepted. This evolution can be detected everywhere. Even powerful forces that tend to try to disrupt harmony out of self-interest gained from chaos face a robust system that can protect itself from their attacks.

7.4.1 Wrapping Up

As a result of this POP exercise centered on air quality, health and lifestyle with civilian involvement and persuasion techniques, we learned a lot. We have tried to express this in this brief. Many people became involved, and they too, in reality, are the co-authors of this work. We write history together as Sustainocracy unfolds itself in front of us in the form of level 4 regional development, featuring much better perspectives for sustainable progress than where we have come from. We learned to put human complexity, personal motivation and reciprocity delivered by the surroundings with which we interact into perspective, whether it is within a money-driven hierarchy or a natural value-driven system. From an anthropological point of view, we see that humankind has the opportunity to evolve from a competitive species that produces its own chaos at regular intervals to a species that generates harmony, cohesion and symbiosis with its environment in the interest of wellness for itself.

The danger we face is that the results of new economic boosts, as the result of focusing on value creation in a self-resilient community, will eventually be taken over by the singular financial hierarchies in an attempt to maintain the status quo of growth. By creating a dual value system, we can counteract this and also establish an economic yin-yang that will keep a watch out for when imbalance is generated. The search for balance will be permanent, while we accept instruments of growth, collapse and renewal as part of our regional eco-system.

In reality, we created a new infrastructure on top of the old mechanisms of regional development. This infrastructure makes visible the invisible and interacts with human motivators rather than hierarchical control mechanisms, releasing new powerful energies for creative power. This resonates with values other than those of economics, producing a new human world with which to interact. We have produced a workable choice between one system that produces sustainable progress and one that does not. All this is available for other regions to use, reducing the vulnerability of communities quickly and developing them into resilient co-creations of wellness and progress.

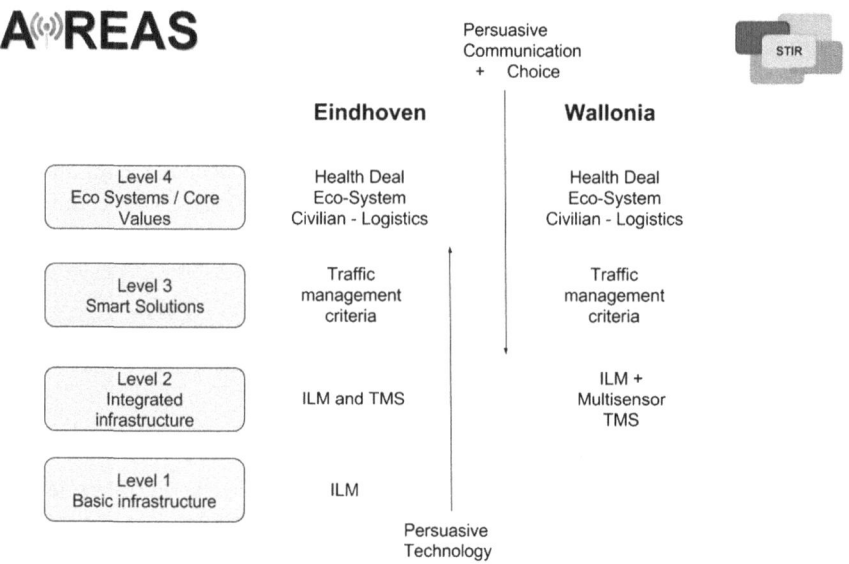

Example how the format extends across regions

7.4.2 Final Conclusion

One of the objectives of the POP was to see if we could create a working format for the purpose of eventually involving 4000 local civilians in a medical investigation. The answer was yes from a working format perspective. However, finding 4000 individuals for medical research proved to be very complicated when dealt with as a fragmented scientific goal. When dealt with within the scope of AiREAS, that of a participation society that needs constant feedback on health and environmental quality in order to make innovative choices, it is, indeed, possible. The research and feedback is then integrated into the overall structure of core values (the Health Deal), with a tremendous amount of sub-objectives marked on its roadmap, each affecting human well-being in one way or another.

Repeating the POP medical and lifestyle research in blocks of 40 participants, connected to each value-driven context along the roadmap, we can reach 4000 and many more for both research and the effort to get all those participants to become pioneers and partners in progressive cultural changes. But before we can do this, a new social and economic commitment needs to arise.

Impact of the POP project on social participation in 2015/2016

Research context	Direct participants (ps + res)	Medical research?	Prime objective	Media Exposure
Axians App	150 + 19	No	Persuasion for expansion	Yes, paper, radio
POP 1	32 + 10	Yes	Medical and Lifestyle	Yes, local paper
Backpack	12 + 11	Yes	Lifestyle and Exposure	Yes, local paper, TV
Marathon	32 + 9	Yes	Persuasion and Medical	Yes, papers, TV, radio
Hackathon	15 + 24	No	Persuasion for expansion and Innovation	No
Erasmus+	120 + 12	No	Persuasion for expansion and Lifestyle	Yes, local papers
Total	361 + 75			>500K local people over 12 times Worldwide: millions of people through open access publications and presentations

The reach of the POP persuasion was much larger than anticipated and much more cost effective as well. We became aware that our exposure to pollution is caused by both lifestyle and our socio-economic context. Our medical research has shown that the negative effects of pollution are largely reversible. Lifestyle can be addressed through awareness programs and persuasive communication. But the socio-economic context requires a commitment from new leadership to put health development before money dependence. The old socio-economic dependence has led many self-aware people to feel blocked in their freedom to address their own health issues due to the economic pressure exerted by the old paradigm. Our POP ends here, with an invitation to the regional, national and global community to embrace their own evolution and accept that value creation and consumption go hand in hand, just like our respiration and heartbeats. If we balance stress and rest in our systems in a natural way, both our economies and our human wellness will be boosted. We will enter a new phase of humankind, one that can last many thousands of years with a stabilized population of around 10–12 billion individuals living in harmony with our planet Earth. The people of this generation will make that difference through the self-aware choice we have. If we don't, we will choke, as nature has taught us. Only health in life survives, always, with or without us.

In the annex, we summarize this POP research as a global recommendation to reach out to each other in a Global Health Deal. The city of Eindhoven and the Province of North Brabant in the Netherlands mark the civilian and directive precedents. This is not a business deal; it is one of shared responsibility, one that can unite us throughout the world, as it has in our small but growing world of AiREAS and sustainocratic processes.

Chapter 8
Annex: Executive Summary and Health Deal

Jean-Paul Close and Eric de Groot

Proof of Principle
Air quality, public health and lifestyle
AiREAS healthy city phases 2 and 3
Executive summary
Global Health Deal
Jean-Paul Close and Eric de Groot
March 2016

8.1 Short Summary

With this summarized research report, AiREAS substantiates the wish and need to establish to a broadly supported Health Deal between the population of any region in the world and its executive governance. Since 2011, AiREAS has become functional as a formal cooperative multidisciplinary workgroup initiated by proactive civilians that invited the participation of local government, innovative entrepreneurship, science and our fellow civilians to create together a healthy city together from a perspective of air quality, public health and regional dynamics. Subsequent research to determine individual exposure to air pollution under the influence of our lifestyle and the reigning cultural economic pressure has shown that both governance (local socio-economic context) and the civilian population (culture and lifestyle) need to

J.-P. Close (✉)
AiREAS, Eindhoven, The Netherlands
e-mail: jp@stadvanmorgen.com

E. de Groot
Imagelabonline and Cardiovascular, Eindhoven and Lunteren,
The Netherlands
e-mail: ericdg@xs4all.nl

© The Author(s) 2016
J.-P. Close (ed.), *AiREAS: Sustainocracy for a Healthy City*,
SpringerBriefs on Case Studies of Sustainable Development,
DOI 10.1007/978-3-319-45620-1_8

take responsibility together for health and a healthy surroundings as a core value for their own sustainable existence. AiREAS substantiates both the imperative need for context and evolutionary changes in lifestyle through a commitment of combined participation, and a practical solution through innovative co-creation, scientific monitoring insights and proven evolutionary methods.

8.2   Introduction

Human health and a healthy environment should be *the* factors steering socio-economic development by government and business. Right now, this is insufficiently the case, with very serious and *in essence avoidable* consequences for the human being and our natural environment, including the sustainability of our (economic) productivity. In fact, at an executive level, policy choices are being made within the context of a largely obsolete socio-economic agenda that sustains or even produces undesirable situations of degraded health. At the same time, both pollution and illness is being economized through unnecessary and expensive bureaucratic systems. We can determine that there is a vicious circle of financial burdens, resulting in the local and global elimination of our environment and health, which has reached unsustainable proportions with extremely destructive consequences for current and future generations.

The need for a drastic review of our socio-economic agenda, in which we primarily consider our own initiative and responsibility as civilians in our own society for our core values, is therefore extremely urgent. Through the Global Health Deal, with the precedent of the regional Brabants Health Deal in the Netherlands, together with the local build-up of expertise through iconic initiatives such as AiREAS and others resonating with health, the executive level is offered a choice: become part of the creative solution or remain among those who cause the problems?

8.3   AiREAS

AiREAS was founded in 2011 in Eindhoven as a core value-driven cooperation of proactive civilians with the involvement of government, innovative business development and socially committed academics from both the technological and human sciences. In our surroundings, issues like living green, space and air quality play an important role in experiencing and safeguarding human wellness and health. AiREAS addresses these socio-economic challenges with the directly measurable goal of creating smart, healthy cities/regions. AiREAS enables this by inviting all regional stakeholders to interact as equals in a multidisciplinary way, from their own sense of responsibility and innovative initiative, to improve environmental conditions for all of us as civilians. Together with her partners, AiREAS has established, in a very short time, a multidisciplinary monitoring system, based on a human values oriented philosophy and by science and policymakers for monitoring environment and health, outside of and within the human being.

8.4 ILM and POP

As a first step prior to the global enlargement of the AiREAS approach for healthy cities, the Innovative Air Measurement (ILM) system was installed in Eindhoven. Through a network of 36 multiple sensor Airboxes, the ILM measures the most important air quality parameters in near real time at the exposure level of citizens. The ILM has now been functioning since the end of 2013, and hence is valid as a method for describing the environment of human beings, linking human wellness to lifestyle and environmental conditions. Validated and broadly accepted parameters for the human context are used interactively between specialists and participants. The perception of wellness is evaluated through extensive interview techniques. The effects in the physical location of participants is traced through GPS while, at the same time, a continuous registration of their heart rhythm and stress takes place. Short- and long-term heart and vascular health are evaluated through blood pressure measurements, vascular functionality and aging registration. Considering integrity and current privacy regulations, the human data is made anonymous for further interpretation.

By making visible the invisible, we can adjust our behavior and policy together, thus stepping outside the vicious circle of destruction, through the development of powerful awareness. The societal purpose of AiREAS and the continuous stream of rationalized, objective, validated and accessible 'Open Data' about wellness and the physical parameters of the human being and our environment are, therefore, complementary and synergetic. They create a very powerful and bonding manner in which to provide us with broad insight and make us aware of issues around our own human living environment, behavior, wellness and health that we can influence. The goals of AiREAS through this approach are tremendously ambitious, recognizing right from the beginning the societal need for a solid philosophical and scientific foundation in order to produce the contextual transformation.

The first application of the AiREAS approach and measurement techniques was tested in a Proof-of-Principle (POP) project within the city quarter of Gestel. First, the significance and feasibility of the project logistics and execution were tested. It was a surprise to see that the combination of the information from the ILM with human data from such a small group of 30 young and elderly participants could already identify the effects of healthy and less healthy lifestyles on wellness and health. In addition, 18 amateur marathon runners from Eindhoven were investigated in the same manner. Both the POP and the data from the marathon project are integrated into Fig. 8.3, in which we describe the vascular age of the different groups. This combined information illustrates that healthy surroundings and life-styles measurably improve health, and that these are factors that can be influenced. All these AiREAS results are based on experience and knowledge of the existing data, providing AiREAS, in furtherance of the societal need, with a solid scientific foundation.

In a few short paragraphs, we will describe the philosophy that we followed, the set-up and, most importantly, the results in the field of human wellness and physical fitness from the Proof-of-Principle. The summary will be finalized with conclusions and recommendations.

The AiREAS Proof-of-Principle (POP) project, approach and its most important results.

8.4.1 History and Purpose of the POP

After rolling out the first phase of AiREAS in Eindhoven, the fine maze mea-surement network ILM measuring the citizenry's exposure to outdoor air pollution, two significant research questions arose:

- How does exposure relate to the development of health and the lifestyles of the citizens? (later indicated as 'AiREAS Phase 2')?
- How can we positively influence this by stimulating innovative measures while making visible the invisible (later indicated as 'AiREAS Phase 3')?

For a truly complete AiREAS project, with an impact on the entire city popu-lation, we estimated that about 4000 people needed to participate (5 % of the local Eindhoven population, or 1 out of 20 civilians). This was also the size of a pop-ulation needed for research on overall wellness and health. This size estimate is based on previous research projects in other cities around the world that studied the relationship between air quality and health parameters.

Due to the size and complexity of such an extensive project, involving 4000 participants, the recent and innovative development of the ILM, our new way of working, a new multidisciplinary team, the inclusion of e/Health research around health and lifestyle, and experimentation with persuasive communication for the purpose of stimulating the innovation impulse, we chose to begin with a smaller test project. This pre-investigation involved 30 people and was characterized as the Proof-of-Principle, or POP, for short. This POP in Gestel later became AiREAS phase 2. The research on 'Persuasive Communication' also became a separate project, referred to as AiREAS phase 3.

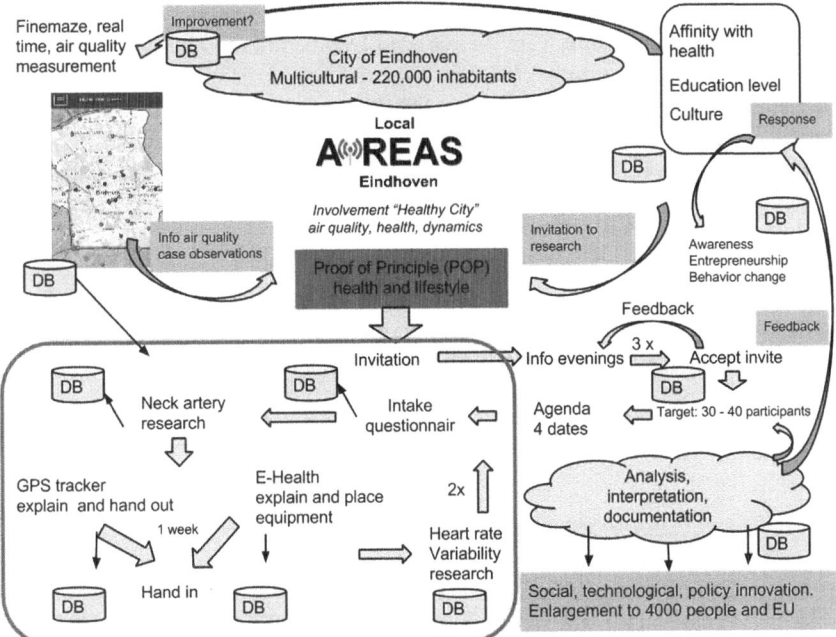

Fig. 8.1 The set-up is based on a broad research project tested by the POP on the complexity of logistics, cooperation and research in civilian participation and science. Process/steps were interactive and quality-controlled (quality control (QC) and quality assurance (QA))

8.4.2 Approach

The civilian participants and the medical research team together formed one community to execute a complex program. Participants were volunteers from the Environmental Defense foundation in Eindhoven, all representing the same affiliation but being of different ages, lives and lifestyle patterns. Figure 8.1 gives an impression of the complexity of the set-up and execution of the POP. Here, it is important to note that each of the actions was realized interactively together with the participants. Technically, all the steps of the process were quality-controlled (QC/QA).

8.4.3 Execution

In January 2015, the research was begun under the overall supervision of Jean-Paul Close. The medical part (heart and vascular research, heart rhythm/stress by means of heart rate variability measurement, eHealth) and the echographic research of arteries and blood pressure were coordinated by Dr. Eric de Groot (ImageOnline) and Dr. Ir. Pierre Cluitmans (TU/e). The civilian participation and persuasive communication were coordinated by Jean-Paul Close, together with Nicolette Meeder, John Schmeitz and Dr. Jaap Ham (TU/e). John Schmeitz was also responsible for the ICT support and data management. Jean-Paul Close furthermore accounted for the local entrepreneurial incubator trials and international enlargement.

The POP set-up produced 11 databases that needed to be combined in order to get a holistic view of the relationship between lifestyle, exposure to air pollution and health. The individual results were reported to each of the participants within the scope of their privacy. The common anonymized insights have been processed into the conclusion of this summary. The specific insights are being prepared for worldwide publication through the scientific publisher Springer.

8.4.4 Financial Investment

The total monetary investment for the POP was 200,000 euro, 75 % of which was financed by the city of Eindhoven and 25 % by the Province of North Brabant. The investment in human participation, knowledge, time and energy of the partners was significantly greater than that.

8.4.5 Most Important Results of the POP, Quality of Life/Wellness and Vascular Age

The results are being published extensively. We highlight the results here of wellness (interview) and measurements of vascular wall thickness (vascular age) in brief. Both wellness and vascular quality are important parameters for health.

8.4.5.1 Interview

The participants in the Gestel POP became partners in the process of healthy city development. It was not just the insight into their health conditions and lifestyles that mattered for the POP, but also the way in which the research and interaction affected their awareness and behavior. The participants were asked to judge their awareness on a scale of 0 (not aware) to 5 (very aware, with adjustment of behavior). The evaluation took place according to the 5 core values defined by the

Fig. 8.2 Wellness and awareness about quality of life before and after taking part in the POP

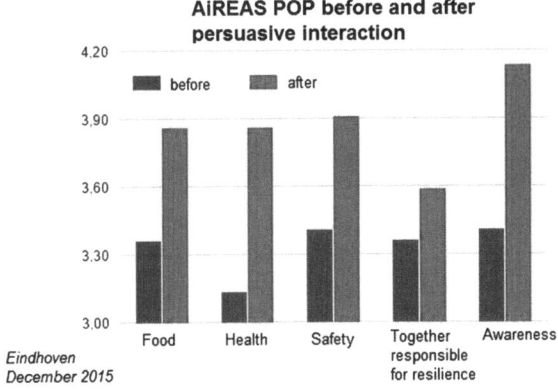

STIR Foundation as the common responsibility between executive governance and the population to assure sustainable continuity. The participants were interviewed twice (by Nicolette Meeder), the first time during the medical research program in March 2015, the second time eight months later upon receipt of their personalized reports in October 2015. In Fig. 8.2, we show the impressive results.

The impressive change in awareness also highlighted a significant problem. Responding participants confirmed that they could address social innovation through their daily decision-making processes. Lifestyle is largely seen as the responsibility of the individual him or herself, often greatly influenced by the surrounding culture. A lot of exposure to pollution is caused by the socio-economic structure of society, in which one needs to participate to sustain an income. Many executive decisions are still made within this old socio-economic context, contributing to and even rewarding polluting and illness-producing behavior. Lifestyle is largely personal, but the socio-economic context is a regional leadership issue. Health hence needs to be a common innovative commitment between leadership and society.

8.4.5.2   Vascular Aging

All participants in the POP took part in the echo research examining the walls of their neck artery (carotids). The thickness of this artery is an accepted measure for the status of health of the vascular system (cardiovascular health): the thinner the wall of the artery, the younger the vascular age and the smaller the risk of heart and vascular diseases (Fig. 8.3).

In addition to the POP, in October 2015, echo-research was also done among (amateur) marathon runners during the Eindhoven Marathon event. These runners

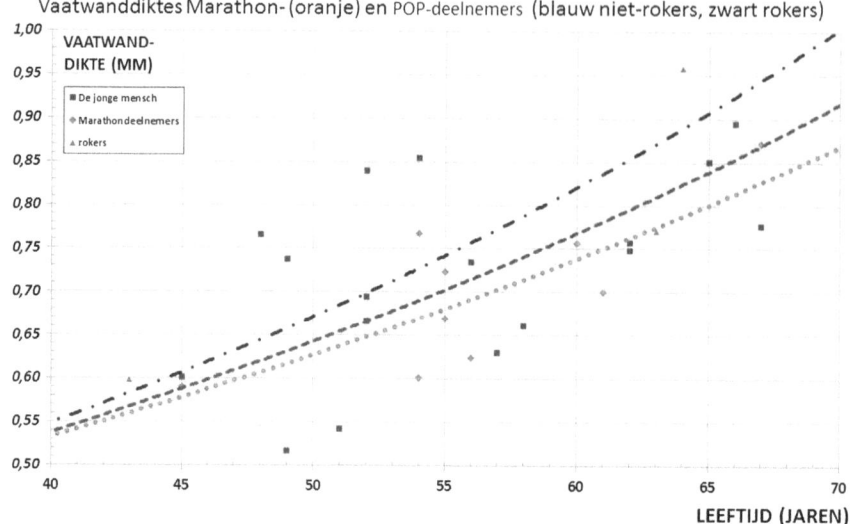

Fig. 8.3 Vascular wall *thickness* versus age. From the POP results, in accordance with previous research, it was determined that people who smoke (=less air quality) show considerably worse arteries compared to non-smokers (the *black dotted line* in the graphic are people who smoke). *Orange* represents the marathon runners, *blue* the other POP participants

had a significantly better vascular age (orange dotted line in the graphic). The blue dotted line shows the vascular situation of the healthy, non-smoking POP partici-pants who had no intensive activity in sports. On a biological level and in terms of risk, the changes caused by smoking and air pollution are tremendous. A healthier lifestyle and a healthier environment therefore lead to healthier organs.

8.5   Schematic Summary, Conclusions and Recommendation

The multi-faceted impact of the AiREAS projects is summarized in the following table.

Impact of the POP project on social participation in 2015/2016

Research context	Direct local participants ~ researchers (ps + res)	Medical research	Primary purpose	Media exposure
Axians App on telephone	150 + 19	No	Persuasive communication	Yes, newspaper, radio, worldwide publication
POP 1	32 + 10	Yes	Medical and lifestyle	

(continued)

(continued)

Research context	Direct local participants ~ researchers (ps + res)	Medical research	Primary purpose	Media exposure
				Yes, newspaper, worldwide publication
Backpack	12 + 11	Yes	Lifestyle and exposure	Yes, newspaper, TV
Marathon	32 + 9	Yes	Medical and sport, communication, lifestyle, incubator	Yes, newspaper, TV, radio
Hackathon	15 + 24	No	Incubator and innovation	No
Erasmus+	120 + 12 (+700 studenten Turkije)	No	Incubator multi-culture and lifestyle	Yes, newspaper
Total involvement and exposure	361 + 75 (+700)			Local >500 K citizens Up to 12 times Worldwide: many millions via Springer and New Horizon publications Presentations

8.6   Reasoning from a New Socio-economic Context: Health

Because all partners in AiREAS already reason from the position of the new socio-economic context based on core value-driven innovation around health, proof has been created of the local impact of such an intense psycho-social transition and awareness in all areas and at all levels of society, individually, human, societal, ecological and economic (hence also referred to as 4× profit). The POP impact after just one year was gigantic. But AiREAS is just an island in an ocean of old socio-economic influence. If the latter also begins to consider the transition as a step-by-step evolution, then an acceleration will develop on the human, social, ecological *and economic* scales due to the intense push of transformative innovation. That is why AiREAS supports building up towards a general public/executive Health Deal in Brabant and proposes extending this worldwide through regional partners.

8.7   More Proof and Results

The first phase of the POP was a challenge in the field of multidisciplinary cooperation between a large diversity of expertise. Even though the insights gathered were successful and multitudinous, they still offered a rather narrow insight into the complexity of the entire city. Thanks to the composition of the POP team, we were able to determine different test areas through which a broader view could be obtained on the reality of the functioning of the entire social community of the city in relation to air pollution, human exposure and behavior via lifestyle. Each person is different, and on top of that, Eindhoven is very multicultural. This segregation in society demanded special dynamics and adaptability from the POP team that only became possible after first working out the medical rounds. Different sub-contexts were necessary to confirm our insights or complement them. In this way, we were able to establish the following research relationships within the same overall context of health and air quality:

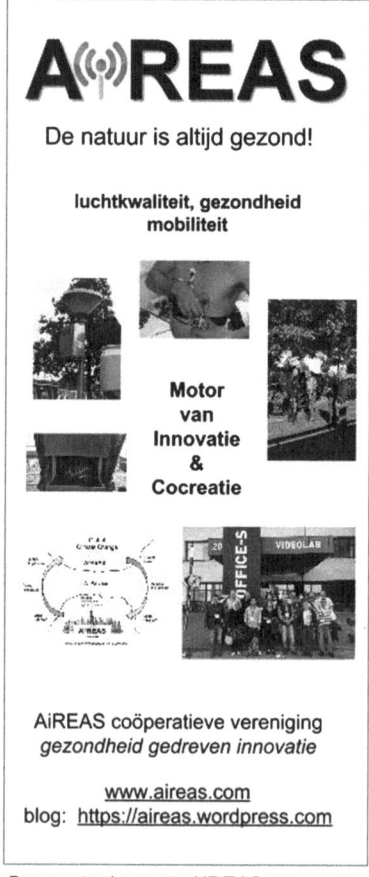

Banner to decorate AiREAS encounters

- Health, outdoor lifestyle and membership Environmental Defense foundation (mindset)/POP1
- Awareness in relation to social innovation/POP2
- Health, direct exposure indoors and outdoors, lifestyle/Backpack
- Turkish community, lifestyle and awareness/Erasmus+
- Health, awareness, technology and entrepreneurship/Hackathon
- Health, sport, air quality and lifestyle/Marathon
- Health, sport and media/Marathon
- Air quality, lifestyle and socio-economic context/All.

All these relationships could be made by looking at the many societal sub-contexts, investigating them and comparing behavioral patterns. Thus, the insight developed that our current socio-economic mainstream culture stand structurally in the way of health by rewarding and valuing the wrong behavior, after which illness and societal costs grow.

By addressing the groups from an alternate socio-economic context, we could make visible the invisible and also show the positive effects if we were able to formalize the health context as the norm. The impact on societal harmony, cohesion and productivity proved to be enormous.

The Health Deal in Brabant and worldwide is hence, according to AiREAS, not a choice but an evolutionary step that is already visible in many awareness-driven, pioneering initiatives throughout society. It only lacks the overall coverage and facilitation of a governing framework. With the Health Deal, this unprecedented evolutionary step will be confirmed and enhanced on the executive level.

8.8 Conclusion

The ILM, POP and Marathon projects illustrate the negative effects on health of our environment and lifestyles, but at the same time, show that they can be resolved. They illustrate above all the enormous potential of working together on innovation in a multidisciplinary setting, and how innovative smart city initiatives such as AiREAS can help with that through the participation of society around core values such as Health.

As indicated in the introduction, this requires modifications in both our lifestyle and the socio-economic agenda of civilians, government and business. These modifications and corresponding changes in insight and the method for making policy is a necessity. We can deal with the modification and adjustment to the new circumstances when considering *health proactively, making leading in our choices* and integral to the societal context (i.e., not reactive to the lack of health via an unaffordable system of monetary dependence).

The proficiency tests done in the POP and Marathon projects had a very positive effect on the response of participants, their contribution to their own improved quality of life and even the appearance of incubators of new entrepreneurship. Lifestyle is in the hands of the citizens involved, but the socio-economic culture is not; that is an executive leadership issue. The latter still rewards and stimulates unhealthy situations, behavior and structures. The reward system is used to sustain a living standard (housing, food supply, social securities, health care). To break this, a socio-cultural turnaround is needed that can only be done by regional executives and citizens _together_. This has the following effects, already visible in the POP:

- The societal context changes health results into totally new entrepreneurial dynamics with health-driven productivity, serviceability and even new desires for landscape and city design.
- The psychosocial awareness attached to health demands a modified structuring of society and reward systems around value creation.
- By establishing health as the steering mechanism, a totally new governmental charter appears, including in regard to decision-making processes. In Eindhoven, we saw the results of a strengthened social cohesion, reconfirmed leadership and a generalized entrepreneurial spirit.
- Through these choices, icon projects of integral innovation appear. AiREAS is already recognized as an icon of value-driven co-creation. More are being developed.
- The proven multidisciplinary co-creation of our common core values delivers multiple forms of reciprocity without precedent:

  - _Government_: structural cost reduction, reduced bureaucracy and the speeding up of complex processes. Consequences were considered costs, while value-driven co-creation is an investment with a diversity of results. Icon-projects that become visible inspire the world. Every participating region is demographically different, producing unique innovations for itself and the world.
  - _Education_: participative learning models with self-leadership and dynamic clustering around value-driven projects,
  - _Entrepreneurship_: intense innovative stimulus for new product market combinations, design, services, landscapes, city development and cooperative structures. A new socio-economic landscape reveals itself with tremendous local and global growth potential, also referred to as the new Kondratiev wave.
  - _Environment_: social innovation takes care of a new economic impulse, new entrepreneurship develops with alternative value systems. People meet and cluster in health-driven cohesion. Nature regains its important status in our awareness and our relationship with our basic needs (food, energy, water, air).

Health is seen by the money-driven hierarchy as soft, abstract and unrealistic. The POP proves the contrary. It is real, responsible and constructively innovative. The value-driven societal context is a valid successor to the old industrial and consumer economic reality.

8.9 Worldwide Attention

During the period of the POP, AiREAS was approached and visited by China, India, Turkey and different European cities and regions with positive intentions to apply the insights to their own local challenges as well. It was also confirmed by major economic players that the regions of Brabant and Eindhoven are especially attractive for investment because of our focus on health. AiREAS's method of working has become an icon on its own and was recognized through a European innovation award, as well as formal scientific positioning as a peer 4 regional development by an external research company.

Global enlargement is only possible in regions where health is embraced as a core value for regional development. The Brabant Health Deal is a source of inspiration for all regions of the world and the establishment of a local AiREAS would be a proven manner for addressing the transition between socio-economic contexts in a satisfactory way. The STIR Foundation is set up to help all regions of the world make such a commitment and take these important evolutionary steps.

Signed in Eindhoven, March 15th, 2016

Jean-Paul Close   Dr. Eric de Groot
Founder STIR   Heart and vascular research
Researcher and initiator of AiREAS ImageOnline
Dr. Ir. Pierre Cluitmans   John Schmeitz
Eindhoven University of Technology Human being and technology
Heart Rate Variability and stress research   ICT and database

And the rest of the POP team and participants.

Publications:

Springer: AiREAS and Spirituality in Business Ethics https://books.google.nl/books?id=6M6lBQAAQBAJ&lpg=PA81&ots=zwT5BDGykE&dq=Springer%20Jean-Paul%20Close&hl=nl&pg=PA1#v=onepage&q&f=false

New Horizons: Redefining human complexities https://marktleiderschap.wordpress.com/2015/10/10/my-contribution-to-anthropology-and-sociology/

Springer/Phase 1 AiREAS: making visible the invisible http://link.springer.com/book/10.1007/978-3-319-26940-5

Vinci energy Paris: collective intelligence supporting urban air quality https://www.vinci.com/vinci.nsf/en/newsroom/pages/collective_intelligence_supporting_urban_air_quality.htm

Index

© The Author(s) 2016
J.-P. Close (ed.), *AiREAS: Sustainocracy for a Healthy City*,
SpringerBriefs on Case Studies of Sustainable Development,
DOI 10.1007/978-3-319-45620-1